Drilling Engineering

Drilling Engineering

Editor

Manoj Karkare

Drilling Engineering

Edited by **Manoj Karkare**

Printed in 2017

ISBN: 978-1-68117-355-9

Library of Congress Control Number: 2015941547

© 2016 by
SCITUS Academics LLC,
616, Corporate Way, Suite 2, 4766,
Valley Cottage, NY 10989

www.scitusacademics.com

This book contains information obtained from highly regarded resources. Copyright for individual articles remains with the authors as indicated. All chapters are distributed under the terms of the Creative Commons Attribution License, which permits unrestricted use, distribution, and reproduction in any medium, provided the original author and source are credited.

Notice

Reasonable efforts have been made to publish reliable data and views articulated in the chapters are those of the individual contributors, and not necessarily those of the editors or publishers. Editors or publishers are not responsible for the accuracy of the information in the published chapters or consequences of their use. The publisher believes no responsibility for any damage or grievance to the persons or property arising out of the use of any materials, instructions, methods or thoughts in the book. The editors and the publisher have attempted to trace the copyright holders of all material reproduced in this publication and apologize to copyright holders if permission has not been obtained. If any copyright holder has not been acknowledged, please write to us so we may rectify.

Contents

Preface vii

Chapter 1 CFD Method for Predicting Annular Pressure Losses and
Cuttings Concentration in Eccentric Horizontal Wells 1
Titus N. Ofei, Sonny Irawan, and William Pao

Chapter 2 Decorating and Filling of Multi-walled Carbon
Nanotubes with TiO_2 Nanoparticles via Wet Chemical Method 39
Sedigheh Abbasi, Seyed Mojtaba Zebarjad, and
Seyed Hossein Noie Baghban

Chapter 3 Study of Sodium-Chromium-Iron-Phosphate Glass by
XRD, IR, Chemical Durability and SEM 55
Youssef Makhkhas, Said Aqdim, and El Hassan Sayouty

Chapter 4 A Comparison on Core Drilling of Silicon Carbide and
Alumina Engineering Ceramics with Mono-layer Brazed
Diamond Tool using Surfactant as Coolant 71
F.L. Zhang, P. Liu, L.P. Nie, Y.M. Zhou, H.P. Huang,
S.H. Wu, and H.T. Lin

Chapter 5 Nature of Drilling Forces During Spark
Assisted Chemical Engraving 91
Jana D. Abou Ziki and Rolf Wüthrich

Chapter 6 Complexity in Semiconductor Manufacturing, Activity of
Antimicrobial Agents, and Drilling of Hydrocarbon Wells:
Common Themes and Case Studies 101
Michael Nikolaou, Pratik Misra, Vincent H. Tam,
and Andrew D. Bailey III

Chapter 7	Small-Hole Drilling in Engineering Plastics Sheet and its Accuracy Estimation 171
	Hiroki Endo and Etsuo Marui
Chapter 8	Study on the Diamond Tool Drilling of Engineering Ceramics 187
	Q.H. Zhang, J.H. Zhanga, D.M. Sun, and G.D. Wang
Chapter 9	Mathematical Model of Dissolution of Particles of NaCl in Well Drilling: Determination of Mass Transfer Convective Coefficient 201
	L.A. Calcada, L.A.A. Martins, C.M. Scheid, S.C. Magalhães, and A.L. Martins
Chapter 10	Experimental Investigation of the Effect of Machining Parameters on the Surface Roughness and the Formation of Built Up Edge (BUE) in the Drilling of Al 5005 225
	Erkan Bahçe and Cihan Ozel

Citations.. 243

Index.. 247

Preface

Drilling engineering is a subset of petroleum engineering. Drilling engineers design and implement procedures to drill wells as safely and economically as possible. They work closely with the drilling contractor, service contractors, and compliance personnel, as well as with geologists and other technical specialists. The drilling engineer has the responsibility for ensuring that costs are minimized while getting information to evaluate the formations penetrated, protecting the health and safety of workers and other personnel, and protecting the environment.

Editor

Chapter 1

CFD Method for Predicting Annular Pressure Losses and Cuttings Concentration in Eccentric Horizontal Wells

Titus N. Ofei[1], Sonny Irawan[1], and William Pao[2]

[1]Petroleum Engineering Department, Universiti Teknologi PETRONAS, Bandar Seri Iskandar, 31750 Tronoh, Malaysia
[2]Mechanical Engineering Department, Universiti Teknologi PETRONAS, Bandar Seri Iskandar, 31750 Tronoh, Malaysia

ABSTRACT

In oil and gas drilling operations, predictions of pressure losses and cuttings concentration in the annulus are very complex due to the combination of interacting drilling parameters. Past studies have proposed many empirical correlations to estimate pressure losses and cuttings concentration. However, these developed correlations are limited to their experimental data range and setup, and hence, they cannot be applicable to all cases. CFD methods have the advantages of handling complex multiphase flow problems, as well as, an unlimited number of physical and operational conditions. The present study employs the inhomogeneous (Eulerian-Eulerian) model to simulate a two-phase solid-fluid flow and predict pressure losses and cuttings concentration in eccentric horizontal annuli as a function of varying drilling parameters: fluid velocity, diameter ratio (ratio of inner pipe diameter to outer pipe diameter), inner pipe rotation speed, and fluid type. Experimental data for pressure losses and cuttings concentration from previous literature compared very well with simulation data, confirming the validity of the current model. The study shows how reliable CFD methods can replicate the actual, yet complex oil and gas drilling operations.

INTRODUCTION

Predictions of pressure losses and cuttings concentration in annular wells are strongly affected by varying drilling parameters such as fluid velocity, fluid properties (density, viscosity), cuttings size and density, hole-pipe eccentricity, drill pipe rotation, and annular diameter ratios. There are few attempts made by some investigators to estimate pressure losses and cuttings concentration in annular geometries with and without drill pipe rotation by employing either experimental or numerical approaches.

Among the first authors to conduct extensive experimental study on cuttings transport at varying angles of inclinations is Tomren et al. [1]. The authors studied the effects of fluid velocity, fluid rheological properties, pipe-hole eccentricity, drill pipe rotation, and flow regimes on cuttings concentration at steady state condition. They concluded that fluid velocity, hole inclination, and mud rheological properties were the major factors affecting mud carrying capacity. Becker and Azar [2] also investigated experimentally the effects of mud weight and annular diameter ratio on the performance of hole cleaning in inclined wellbores. The authors observed that variations in the drill pipe have minimum effect on particle concentration for the same fluid velocity. According to Adari et al. [3], the practical use of drilling factors in controlling cuttings transport is much dependent on their controllability in the field. It is believed that, cuttings transported in the annulus are not always affected by a single parameter, but a combination of parameters to ensure efficient hole cleaning [4]. Other studies have also confirmed that increase in fluid velocity results in a decrease in cuttings accumulation in the wellbore [5–7]. Ozbayoglu and Sorgun [8] also conducted cuttings transport experiment and developed empirical correlations for estimating pressure losses with the presence of cuttings and drill pipe rotation in horizontal and inclined wellbores. They observed that the influence of drill pipe rotation on pressure loss is more significant if fluid is more non-Newtonian. The annular test section has diameter ratio of 0.62. Similar cuttings transport experiment study was carried out by Sorgun et al. [9] in horizontal and inclined annular geometry of diameter ratio of 0.62. The authors observed that, the existence of cuttings in the system caused an increase in pressure loss due to a decrease in flow area inside the annular gap. Further observation was that, drill pipe rotation decreases the pressure loss significantly if the drill pipe is making orbital motion in eccentric annulus. Another experimental study was conducted [10] to analyse the effects of some "very difficult to identify" data on the estimation of total pressure loss and cuttings concentration in horizontal and inclined annulus. Results from this study indicate that drill pipe rotation does not have significant influence on pressure loss for constant rate of penetration (ROP) and fluid velocity. The annular test section has diameter ratio of 0.64.

One of the pioneering works by Markatos et al. [11] modelled single phase Newtonian flow in nonuniform narrow annular gaps using finite difference technique. The velocity flow fields as well as static pressure were predicted in a two-dimensional flow.

Han et al. [12] is among the first to conduct experimental and CFD studies on solid-fluid mixture flow in vertical and highly deviated slim hole annulus. They concluded that, annular pressure losses increase with mixture fluid velocity, annular angle of inclination, and drill pipe rotation speed. The annular test section has diameter ratio of 0.70. Mokhtari et al. [13] employed CFD method to model the effects of eccentricity and flow behaviour index on annular pressure loss and velocity profile for varying diameter ratios from 0.30 to 0.90. The authors, however, did not include cuttings in the annular mainstream. Recently, Ofei et al. [14] also employed CFD technique to analyse the influence of diameter ratio, fluid velocity, fluid type, fluid rheology, and drill pipe rotation speed on pressure loss in eccentric horizontal wellbore with the presence of cuttings.

The present study also utilises a CFD technique to examine the effects of fluid velocity, annular diameter ratio (ranging from 0.64 to 0.90), drill pipe rotation, and fluid type on the prediction of pressure loss and cuttings concentration for solid-fluid flow in eccentric horizontal wellbore. Contours of cuttings volume fraction, 3D cuttings velocity streamlines and radial cuttings velocity profiles are also presented to give further insight on cuttings transport. The new findings from this study would provide better understanding and guide in the selection of optimum drilling parameters in narrow annuli drilling such as casing drilling and slim holes.

MATERIALS AND METHODS

Multiphase component of CFD software ANSYS-CFX 14.0 is adopted in this study. In ANSYS-CFX, a multiphase flow containing dispersed particles may be modelled using either Lagrangian Particle Tracking model or Eulerian-Eulerian model. The inhomogeneous (Eulerian-Eulerian) model, sometimes called the two-fluid model, regards both continuous and dispersed phases as continuous media. In this study, the Eulerian-Eulerian model is preferred to the Lagrangian Particle Tracking model due to its ability to handle high solid volume fractions. Furthermore, it accounts for solid particle-particle interaction and includes turbulence automatically [15]. A drawback of this model is, however, that they need complex closure relations. The following continuity and momentum equations representing the two-phase flow model are described for the sake of brevity.

Continuity Equations

The fluid phase continuity equation assuming isothermal flow condition can be expressed as [16, 17]

$$\frac{\partial}{\partial t}(\hbar_l) + \nabla \cdot (\hbar_l U_l) = 0. \tag{1}$$

Similarly, for a solid phase,

$$\frac{\partial}{\partial t}(\hbar_s) + \nabla \cdot (\hbar_s U_s) = 0, \tag{2}$$

where the solid and fluid phase volume fraction sum up as follows:

$$\hbar_s + \hbar_l = 1. \tag{3}$$

At steady state condition, $\partial/\partial t = 0$

Momentum Equations

The forces acting on each phase and interphase momentum transfer term that models the interaction between each phase are given below [16, 17].

For fluid phase,

$$\rho_l \hbar_l \left[\frac{\partial U_l}{\partial t} + U_l \cdot \nabla U_l \right] = -\hbar_l \nabla p + \hbar_l \nabla \cdot \bar{\bar{\tau}}_l + \hbar_l \rho_l g - M. \tag{4}$$

Similarly, for solid phase,

$$\rho_s \hbar_s \left[\frac{\partial U_s}{\partial t} + U_s \cdot \nabla U_s \right]$$
$$= -\hbar_s \nabla p + \hbar_s \nabla \cdot \bar{\bar{\tau}}_l + \nabla \cdot \bar{\bar{\tau}}_s - \nabla P_s + \hbar_s \rho_s g + M. \tag{5}$$

At steady state condition, $\partial/\partial t = 0$.

Other Closure Models

Interphase Drag Force Model

For spherical particles, the drag force per unit volume is given as

$$M_d = \frac{3C_D}{4d_s} k_s \rho_l |U_s - U_l|(U_s - U_l). \quad (6)$$

For densely distributed solid particles, where the solid volume fraction, $k_s < 0.2$ the Wen and Yu [18] drag coefficient model may be utilised. This model is modified and implemented in ANSYS-CFX to ensure the correct limiting behaviour in the inertial regime as

$$C_D = k_l^{-1.65} \max\left[\frac{24}{N'_{Re_p}}\left(1 + 0.15 N'^{0.687}_{Re_p}\right), 0.44\right], \quad (7)$$

Where $N'_{Re_p} = k_l N_{Re_p}$ and $N_{Re_p} = \rho_l |U_l - U_s| d_s / \mu_l.$

For large solid volume fraction, $k_s > 0.2$ the Gidaspow drag model may be used with the interphase drag force per unit volume defined as [19]

$$M_D = \frac{150(1 - k_l)^2 \mu_l}{k_l d_s^2} + \frac{7}{4}\frac{(1 - k_l)\rho_l |U_l - U_s|}{d_s}. \quad (8)$$

In this study, both Wen and Yu and Gidaspow drag models were employed depending on the computed solid volume fraction. An approximate method for computing the solid volume fraction is presented in (18).

Lift Force Model

For spherical solid particles, ANSYS-CFX employs the Saffman and Mei lift force model as

$$M_L = \frac{3}{2\pi} \frac{\sqrt{v_l}}{d_s \sqrt{|\nabla \times U_l|}} C'_L \rho_s \rho_l (U_s - U_l) \times (\nabla \times U_l + 2\Omega). \quad (9)$$

Saffman [20, 21] correlated the lift force for low Reynolds number past a spherical solid particle where $C'_L = 6.46$, and $0 \leq N_{Re_p} \leq N_{Re_\omega} \leq 1$. For higher range of solid particle Reynolds number, Saffman's correlation was generalised by Mei and Klausner [22] as follows:

$$C'_L = \begin{cases} 6.46 \cdot f\left(N_{Re_p}, N_{Re_\omega}\right) & \text{for: } N_{Re_p} < 40 \\ 6.46 \cdot 0.0524 \cdot \left(\beta N_{Re_p}\right)^{1/2} & \text{for: } 40 < N_{Re_p} < 100, \end{cases} \quad (10)$$

where $\beta = 0.5\, 0.5(N_{Re_\omega}/N_{Re_p})$,

$$f\left(N_{Re_p}, N_{Re_\omega}\right) = \left(1 - 0.3314\beta^{0.5}\right) \cdot e^{-0.1 N_{Re_p}} + 0.3314\beta^{0.5}, \quad (11)$$

and

$N_{Re_\omega} = \rho_l \omega_l d_s^2 / \mu_l,\ \omega_l = |\nabla \times U_l|$.

Turbulence k-ε Model in Multiphase Flow

The k-ε turbulence model offers a good compromise in terms of accuracy and robustness for general purpose simulations. It is a semiempirical model based on transport equation for the estimation of turbulent length scale and velocity scale from the turbulent kinetic energy (k) and dissipation rate (ε) [23]. In multiphase flow, the transport equations for k and ε are phase dependent and assume a similar form to the singlephase transport equations, respectively, as

$$\frac{\partial}{\partial t}(C_\alpha \rho_\alpha k_\alpha) + \nabla \cdot \left(C_\alpha \left(\rho_\alpha U_\alpha k_\alpha - \left(\mu + \frac{\mu_{t\alpha}}{\sigma_k}\right)\nabla k_\alpha\right)\right)$$
$$= C_\alpha (P_\alpha - \rho_\alpha \varepsilon_\alpha) + T^{(k)}_{\alpha\beta}, \quad (12)$$

$$\frac{\partial}{\partial t}(C_\alpha \rho_\alpha \varepsilon_\alpha) + \nabla \cdot \left(C_\alpha \rho_\alpha U_\alpha \varepsilon_\alpha - \left(\mu + \frac{\mu_{t\alpha}}{\sigma_\varepsilon}\right)\nabla \varepsilon_\alpha\right)$$
$$= C_\alpha \frac{\varepsilon_\alpha}{k_\alpha}(C_{\varepsilon 1} P_\alpha - C_{\varepsilon 2} \rho_\alpha \varepsilon_\alpha) + T_{\alpha\beta}^{(\varepsilon)}. \tag{13}$$

Diffusion of momentum in phase α is governed by an effective viscosity as

$$\mu_{\text{eff}} = \mu + \mu_{t\alpha}. \tag{14}$$

The k-ε model assumes that the turbulence viscosity is linked to the turbulence kinetic energy and dissipation by the following relation:

$$\mu_{t\alpha} = C_\mu \rho_\alpha \frac{k_\alpha^2}{\varepsilon_\alpha}. \tag{15}$$

The governing sets of partial differential equations were discretized using finite volume technique. The discretized equations together with initial and boundary conditions are solved iteratively for each control volume of pressure drop and cuttings concentration using ANSYS CFX 14.0 solver.

Physical Model and Carrier Fluid

Two-phase solid-fluid flow in eccentric horizontal annulus with stationary outer pipe and rotating inner pipe is presented. The inner pipe represents the actual drill pipe while the outer pipe represents the hole. Four annular 3D geometries are modelled with diameter ratios κ = 0.64, 0.70, 0.80 and 0.90 using ANSYS Workbench.

In order to eliminate end effects and ensure fully developed flow, the length of the annular pipe must be longer than the hydrodynamic entrance length. For a single phase Newtonian fluid flowing in a pipe, the hydrodynamic length is presented by Shook and Roco [24] as

$$L_h = 0.062 N_{\text{Re}}(D). \tag{16}$$

However, for a two-phase flow in annular gap with a non-Newtonian fluid, such expression as in (16) does not yet exist in literature. As a rule of thumb, the authors have adopted (16) by replacing the pipe diameter with a hydraulic diameter $D_h = D_2 - D_1$. It should be noted that a much longer annular length would only result in a computationally expensive CFD simulation. Figure 1 shows the physical model for solid-fluid flow. The fluid is considered incompressible, steady state, and isothermal. The rheology of the fluid is described by both Newtonian (water) and Power-Law model. The apparent viscosity for Power-Law model is given as

$$\mu_a = K\dot{\gamma}^{n-1}, \qquad (17)$$

where K is consistency index, $\dot{\gamma}$ is shear rate, and n is flow behaviour index. For $n = 1$, (17) reduces to Newtonian model, $n < 1$, fluid is shear thinning, and $n > 1$, fluid is shear thickening. In this study, $0.3 < n \leq 1$.

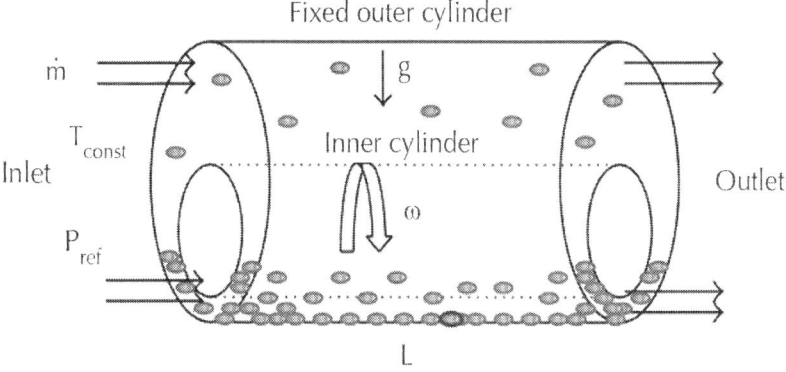

Figure 1: Physical model for solid-fluid flow.

Boundary Conditions and Meshing

A mixture mass flow rate boundary condition was specified at the inlet while zero gauge pressure specified at the outlet. No-slip boundary conditions were imposed on both inner and outer pipe walls for both fluid and particles. The 3D annular geometries were meshed into unstructured tetrahedral grids of approximately 0.66–2.15 × 10^6 elements. Inflation layers were created near the walls covering about 20% of the inner and outer radii for resolving the mesh in the near-wall region as well as accurately capturing the flow effects in that region. Figure 2 shows the 3D section of the meshed annular geometries.

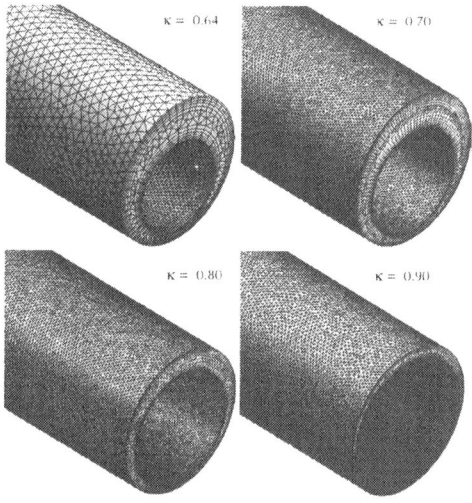

Figure 2: 3D section of meshed annular geometry.

Grid Independence Study

To optimise the mesh sizes until results were insignificantly dependent on mesh size, grid independence study was conducted for all diameter ratios. The carrier fluid used is water flowing at a velocity of 2.743 m/s and the inner pipe rotation speed is 80 rpm. The cuttings feed concentration which gives an idea of the amount of particles in motion that are introduced to the annular space. This is computed as a function of area of bit, fluid velocity, and rate of penetration (ROP) as [6]

$$C_{ct} = \frac{(ROP) A_{bit}}{R_T Q}, \tag{18}$$

where R_T is defined as the ratio of the particle transport velocity to the average annular fluid velocity. For the purpose of this study, R_T is taken as 0.5 based on experimental findings [9]. Figures 3(a) to 3(d) show the variation of pressure losses as a function of element sizes. In Figures 3(a) and 3(b), element size of 0.003 m and below would result in insignificant changes in pressure losses; however, more computational time is required for elements sizes below 0.003 m. In Figures 3(c) and 3(d), element size of 0.003 m and above also shows no significant changes in pressure losses. An optimum element size of 0.003 m is chosen for all diameter ratios resulting in approximately 0.66–2.15 × 10⁶ number of elements with increasing diameter ratio from 0.64 to 0.90. The CPU time recorded in this study ranges between 7.2 × 10³ s to 5.4 × 10⁴ s. The simulations were run on a computer with the following specifications: Windows 7 64-bit operating system, with 4 GB RAM, and Pentium Dual-Core processor at 2.3 GHz.

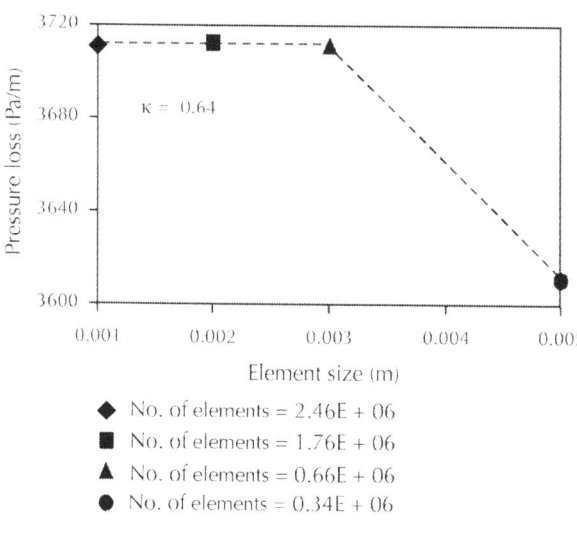

(a)

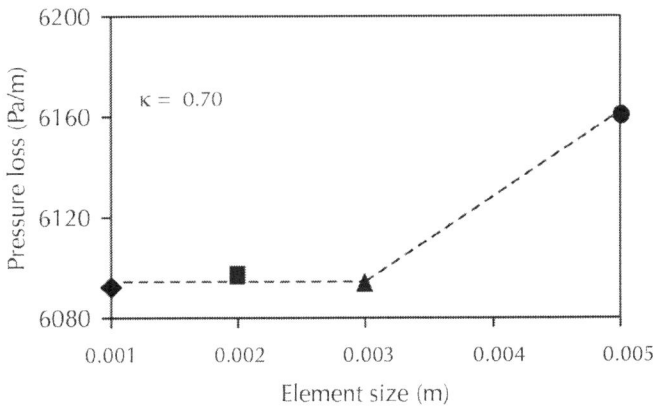

(b)

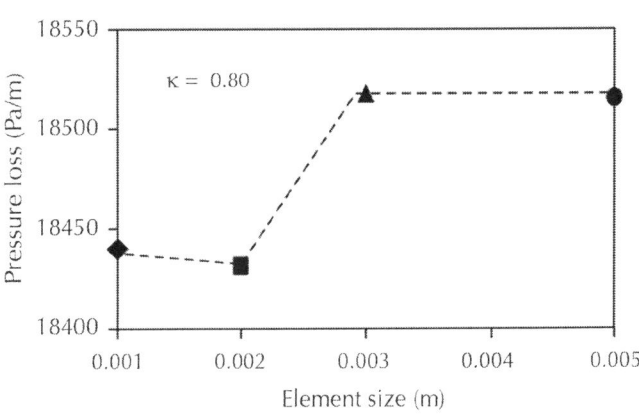

(c)

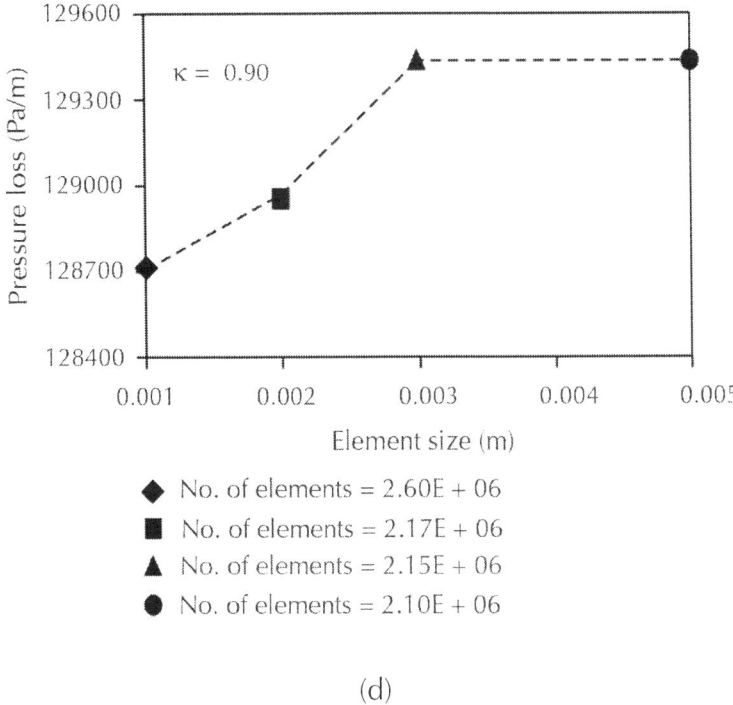

(d)

Figure 3: Grid independence study: (a) $\kappa = 0.64$, (b) $\kappa = 0.70$, (c) $\kappa = 0.80$, (d) $\kappa = 0.90$ at 80 rpm.

Simulation Model Validation

The simulation model setups were validated against experiment data available from previous studies. Pressure loss and cuttings concentration data for cuttings-water flow in a horizontal wellbore were adopted from Osgouei [25]. Also, pressure loss data using non-Newtonian fluid of 0.4% CMC solution for cuttings transport experiment were adopted from Han et al. [12]. Table 1 summarises the rheological properties and operating parameters for the experimental studies.

Table 1: Summary of experimental data from previous studies

Experimental data source	Fluid density (kg/m³)	n(—)	K(Pasⁿ)	Pipe rotation (rpm)	K (D₁/D₂)	e (—)	Fluid velocity (m/s)	Avg. cuttings size (m)	Cuttings density (kg/m³)	ROP (m/s)
Osgouei [25]	998.5	1.0	0.001	80	0.64	0.623	1.524–2.7432	0.00201	2761.4	0.00508
Han et al. [12]	998.5	0.75	0.048	0	0.70	0	0.327–0.654	0.001	2550	0.00526

From Figure 4(a), the calculated pressure loss slightly overpredicted the experimental data by a mean percentage error of 0.84%. Similarly, the calculated cuttings concentration data slightly overpredicted the experimental data by a mean percentage error of 12% as shown in Figure 4(b). The total cuttings concentration, C_{cT} is defined as

$$C_{cT} = \frac{\text{Net volume occupied by particles}}{\text{Total volume of annlulus}} \times 100. \quad (19)$$

Moreover, Figure 4(c) shows the calculated pressure loss deviating slightly from the experimental data by a mean percentage error of 2.5%. The analyses show a good agreement between calculated and experimental data confirming the validity of the current model setup.

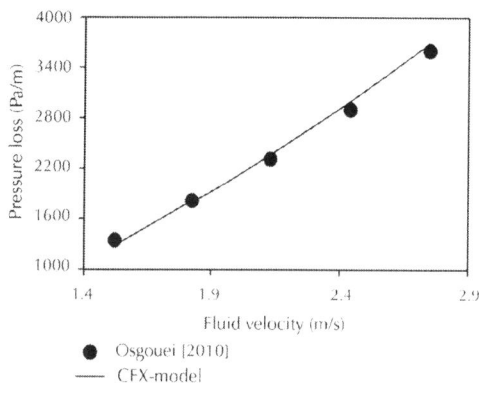

(a)

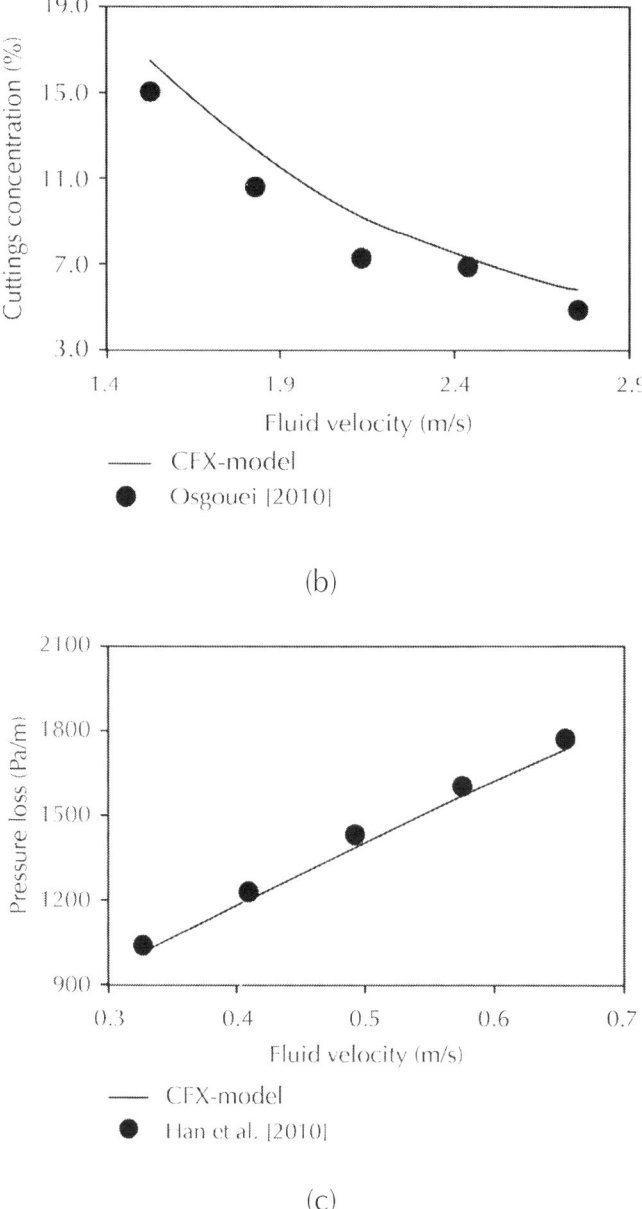

Figure 4: Experimental and simulation data comparison: (a) pressure loss data for cuttings-water flow, (b) cuttings concentration data for cuttings-water flow, (c) pressure loss data for cuttings—0.4% CMC flow.

Simulation Study

Table 2 summarises the simulation setup including fluid rheological properties and drilling parameters. The present study adopts the Eulerian-Eulerian model to simulate a two-phase solid-fluid flow in eccentric horizontal annuli. ANSYS-CFX solver, which is based on a finite volume method [26], is used to solve the continuity and momentum equations with the appropriate initial and boundary conditions. The solution is assumed to be converged when the root mean square (RMS) of the normalised residual error reached 10^{-4} for all simulations. Both Newtonian (water) and non-Newtonian (Power-Law model) fluids are used as carrier fluids. Variations in annular pressure losses and cuttings concentration as a function of fluid velocity, diameter ratio, inner pipe rotation speed, and fluid type are analysed and results are presented. In addition, contours of cuttings volume fraction and cuttings velocities, streamlines of cuttings velocities, as well as profiles of cuttings velocities are also presented.

Table 2: Simulation data for cuttings fluid flow

Rheological and drilling parameter	Case1	Case2
	Water	Mud
Fluid density (kg/m^3)	998.5	1006.3
Cuttings density (kg/m^3)	2761.4	
Avg. cuttings size (m)	0.00201	
Flow behaviour index, n	1	0.51
Viscosity consistency, K, (Pa s^n)	0.001	0.289
Fluid velocity (m/s)	1.524–2.7432	
Inner pipe rotation speed (rpm)	0, 80, 120	
Diameter ratio ($\kappa = D_1/D_2$)	0.64, 0.70, 0.80, 0.90	
Eccentricity (e)	0.623	
ROP (m/s)	0.00508	

RESULTS AND DISCUSSION

Effect of Fluid Velocity

Previous studies [1, 4] have revealed that fluid velocity is a dominant factor during cuttings transport. This phenomenon is also observed in this study. Figure 5 presents the variations in pressure loss and cuttings concentration as a function of increasing annular fluid velocity at constant diameter ratio and 80 rpm. Using both water andmud as carrier fluids, increasing fluid velocity significantly increases pressure losses, while a decrease in cuttings concentration occurs for each constant diameter ratio. This effect is however more pronounced for $\kappa = 0.90$. Figures 5(a)–5(d) depict these observations. For instance, when using mud as carrier fluid in Figure 5(c) and for $\kappa = 0.90$, annular pressure loss was dramatically increased by 97% when the flowing fluid velocity increased from 1.524m/s to 2.749m/s. Similarly, as shown in Figure 5(d), the cuttings concentration decreased by 37% in the annulus as fluid velocity increased from 1.524 m/s to 2.749 m/s for $\kappa = 0.90$. Another observation is that, in Figure 5(b), where the carrier fluid is water, there is almost no variation in cuttings concentration as fluid velocity increases for $\kappa = 0.90$. This indicates that in extreme narrow annuli, lower fluid velocities are capable of transporting enough cuttings fromthe annulus, from1.524m/s to 2.749m/s for $\kappa = 0.90$when the carrier fluid is mud.

18   Drilling Engineering

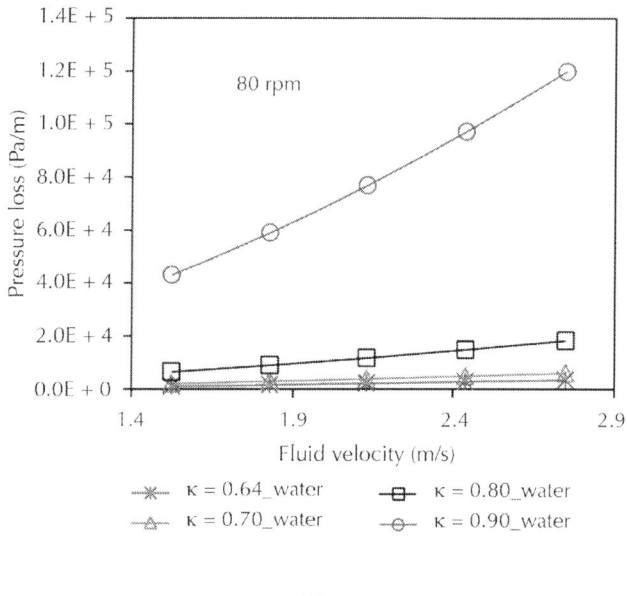

(a)

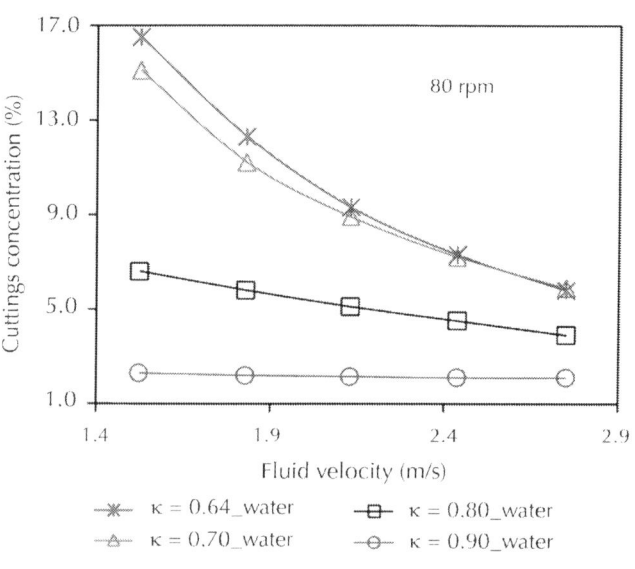

(b)

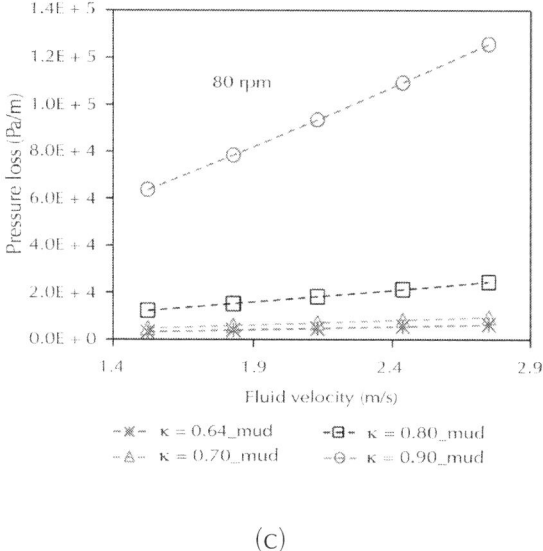

(c)

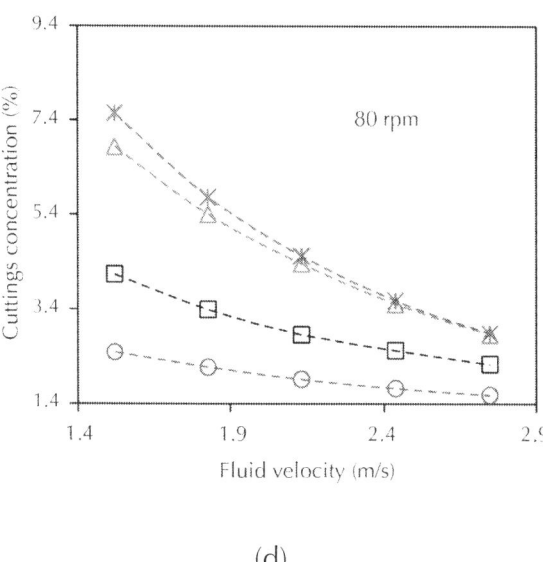

(d)

Figure 5: Effect of fluid velocity at constant diameter ratio on (a) pressure loss with water as carrier fluid, (b) cuttings concentration with water as carrier fluid, (c) pressure loss with mud as carrier fluid, (d) cuttings concentration with mud as carrier fluid.

Effect of Diameter Ratio

Figure 6 presents the influence of diameter ratio on pressure loss and cuttings concentration at constant fluid velocity and 80 rpm. Analyses are shown for both water andmud as carrier fluids. For all cases, as diameter ratio increases from $\kappa = 0.64$ to 0.90, an increase in pressure loss also occurs, whereas a decrease in cuttings concentration is observed for each constant fluid velocity. This influence is however more pronounced for $\kappa = 0.90$. As the annular gap becomes narrower, there are more interactions between cuttings-fluid and pipe walls which results in an increase in friction, and hence, pressure losses. It is worth noting that while the pressure loss difference between $\kappa = 0.64$ and $\kappa = 0.90$ could result in extreme increase by over 3600%, a decrease of about 86%could be realised for cuttings concentration as water flows with a velocity of 1.524 m/s (see Figures 6(a) and 6(b)). Moreover, in Figure 6(b), where the carrier fluid is water, there is almost no disparity in cuttings concentration when $\kappa = 0.90$ for each constant fluid velocity. Although better cuttings transport could be observed in very narrowannuli, optimumdrilling parametersmust be selected to prevent excessive damage to the formation.

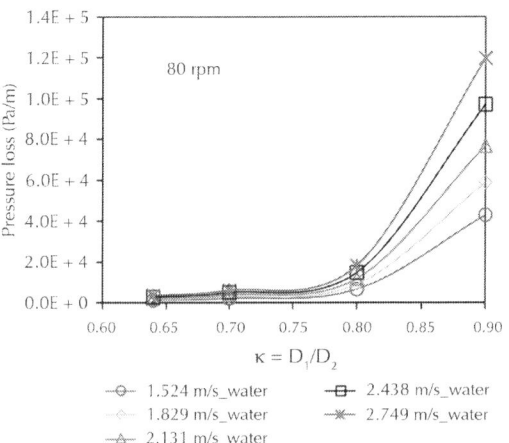

(a)

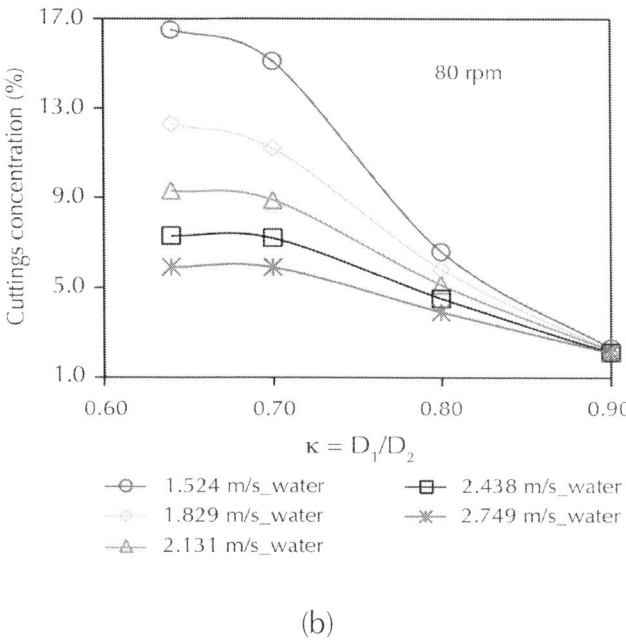

(b)

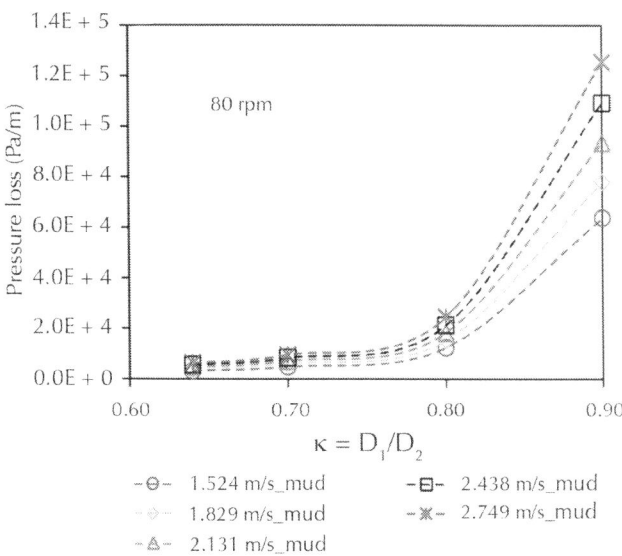

(c)

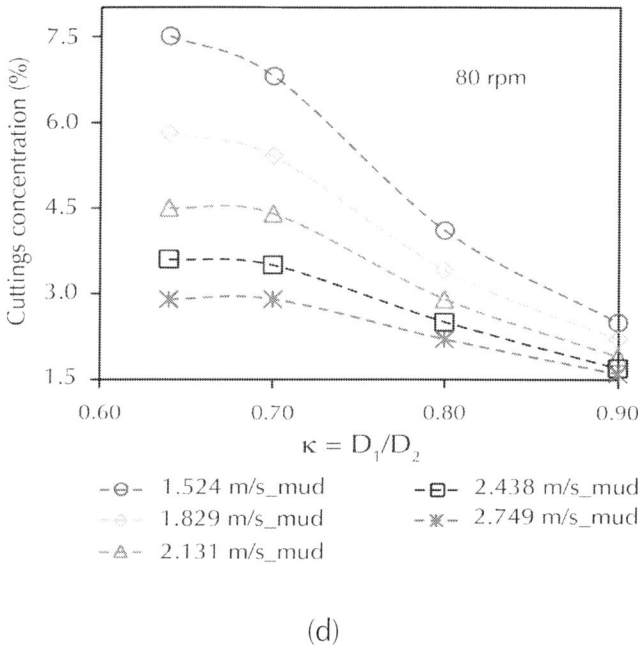

(d)

Figure 6: Effect of diameter ratio at constant fluid velocity on (a) pressure loss with water as carrier fluid, (b) cuttings concentration with water as carrier fluid, (c) pressure loss with mud as carrier fluid, (d) cuttings concentration with mud as carrier fluid.

Effect of Drill Pipe Rotation

The effect of increasing drill pipe rotation on pressure loss and cuttings concentration is shown in Figure 7 when using both water and mud as carrier fluids. In Figures 7(a) and 7(c), an increase in drill pipe rotation speed from 80 rpm to 120 rpm did not result in any significant increment in pressure losses with both water and mud as carrier fluids. The effect on cuttings concentration is quite predominant especially in annular gaps with diameter ratio below $\kappa = 0.70$. For water as carrier fluid, as shown in Figure 7(b), the influence of increasing drill pipe rotation speed from 80 rpm to 120 rpm had a negative impact where the cuttings concentration increased when the diameter ratio is below $\kappa = 0.70$. To explain this behaviour, the low viscous water would generate high turbulence as a function of both axial and rotational flows; which in addition to gravity could cause rapid settling of cuttings in the annulus. Above $\kappa = 0.70$, the influence is virtually the same on cuttings concentration for each constant fluid velocity. On the contrary, when the carrier fluid is mud, as shown in Figure 7(d), increasing drill pipe rotation speed from 80 rpm to 120 rpm shows a decrease in cuttings concentration for a diameter ratio range of $0.64 \leq \kappa < 0.80$. Above $\kappa = 0.80$, the influence is relatively negative on cuttings concentration. In all cases

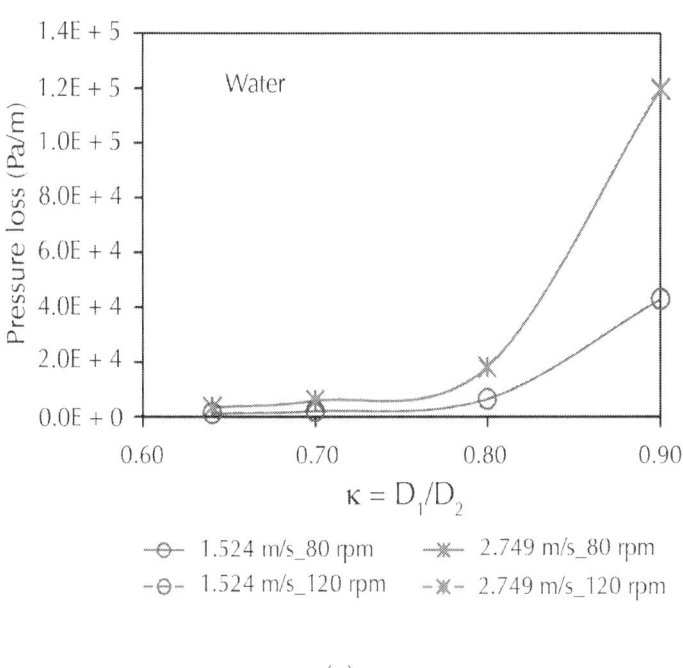

(a)

24 Drilling Engineering

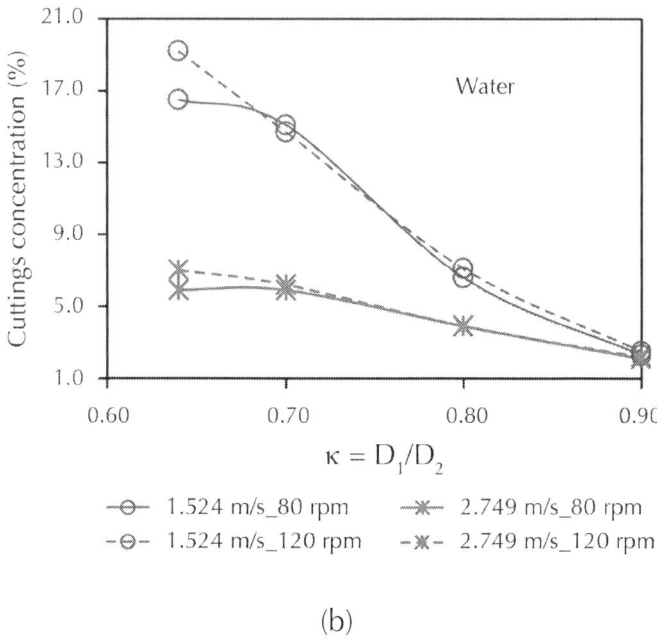

(b)

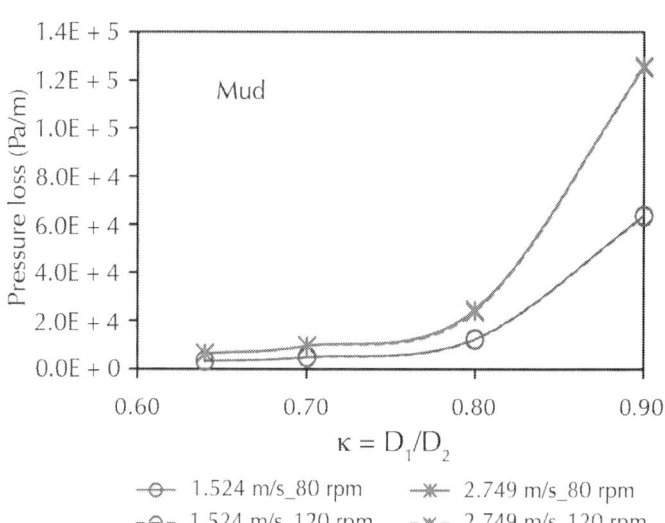

(c)

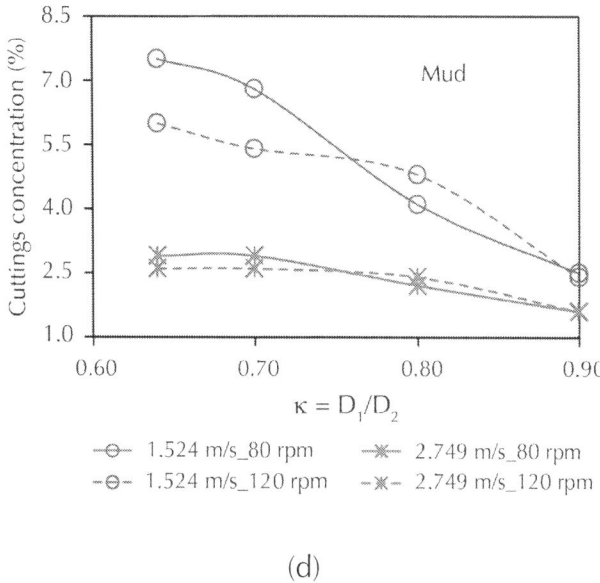

(d)

Figure 7: Effect of drill pipe rotation speed on (a) pressure loss with water as carrier fluid, (b) cuttings concentration with water as carrier fluid, (c) pressure loss with mud as carrier fluid, (d) cuttings concentration with mud as carrier fluid.

Effect of Fluid Type

The effect of Newtonian (water) and non-Newtonian Power-Law fluid (mud) on pressure loss and cuttings concentration are analysed in Figures 8(a) and 8(b), respectively, at 120 rpm. With mud as carrier fluid, high pressure losses were recorded compared to water especially at low fluid velocity and $\kappa = 0.90$ (see Figure 8(a)). Similarly, the mud transported much cuttings compared to water especially at a constant diameter ratio of $\kappa = 0.64$ and low fluid velocities (see Figure 8(b)). For example, 19.2% and 6.0% concentration of cuttings remained in the annulus after flowing with water and mud, respectively, for $\kappa = 0.64$ and fluid velocity of 1.524m/s. The performance of both fluids on cuttings concentration is quite similar at high diameter ratios.

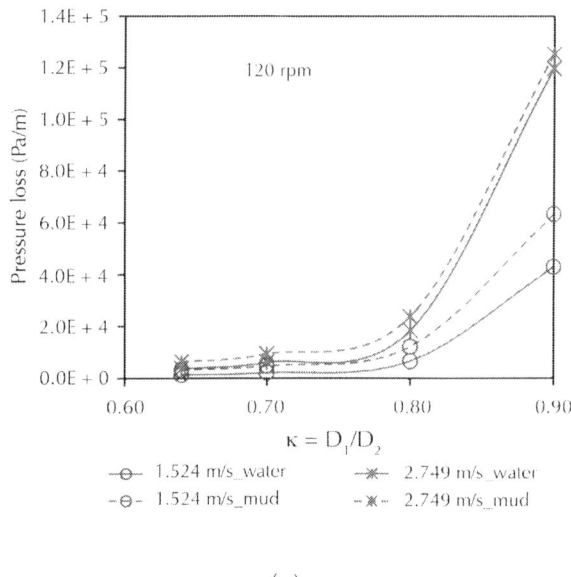

(a)

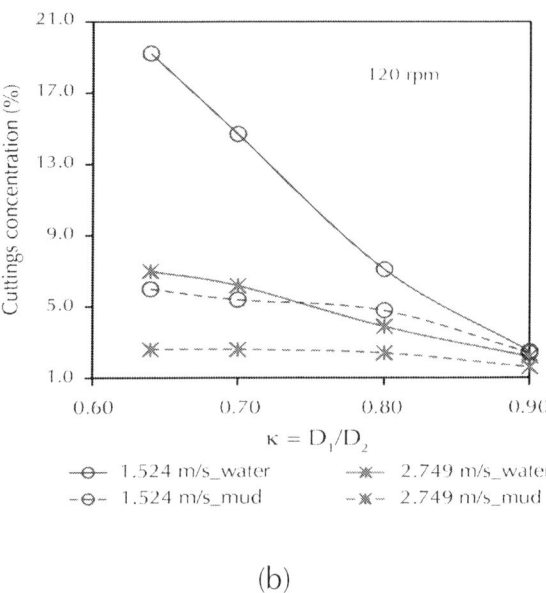

(b)

Figure 8: Effect of fluid type on (a) pressure loss and (b) cuttings concentration.

Cuttings Volume Fraction, Velocity, and Profiles with Water as Carrier Fluid

Figures 9–11 show the contours of cuttings volume fraction, 3D streamlines of cuttings velocities, and radial measurements of cuttings velocity profiles, respectively, flowing with water at 1.524 m/s. As shown in Figure 9, the cuttings concentration accumulates in the narrowest gap of the eccentric annuli forming a bed due to gravity and the low viscosity of the carrier fluid. However, the rotation of the drill pipe from 0 rpm to 120 rpm reduces the cuttings bed by sweeping it into the widest gap where the fluid velocity is high to transport them to the surface. This observation is evident for all diameter ratios and shows the significance of drill pipe rotation in minimising differential pipe sticking, cuttings bed erosion, as well as excessive pressure losses. Figure 10 also depicts 3D streamlines of cuttings velocity. From the colour legend, the velocity of cuttings is high at some distance from the annular inlet and decreases to a minimum velocity towards the exit of the annular geometries. The decrease in cuttings velocity is an indication of cuttings settling to form a bed due to the low viscous nature of the carrier fluid and gravity. Drill pipe rotation induces a rotational flow on the cuttings bed into the annular mainstream and carries them to the surface. This rotation effect reduces the annular bed area for all diameter ratios. The radial measurements of cuttings velocity profiles at 1.524 m/s and 120 rpm are also presented in Figure 11. The radial distance is normalised. In the widest gap of the annular area, as shown in Figure 11(a), cuttings velocity increases with increasing diameter ratio where the peak velocities calculated are 1.896 m/s, 1.970 m/s, 2.043 m/s, and 1.999 m/s for k=0.64, 0.70, 0.80 and 0.90, respectively. On the contrary, in Figure 11(b), the cuttings velocity in the narrowest annular gap show irregular profiles as diameter ratio increases. The effect of drill pipe rotation is seen to have greater impact on the cuttings velocity especially near the vicinity of the drill pipe where there is high shear. For example, at $\kappa = 0.90$, the peak cuttings velocity recorded was 0.481 m/s, and it occurred at the vicinity of the drill pipe.

28   Drilling Engineering

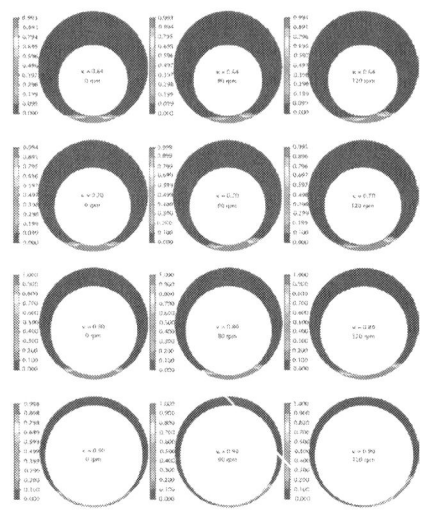

Figure 9: Contours of cuttings volume fraction for varying diameter ratios and inner pipe rotation with water as carrier fluid at 1.524 m/s.

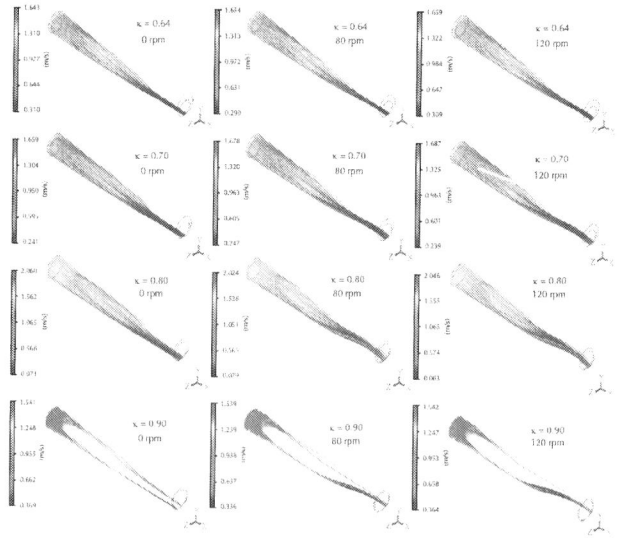

Figure 10: 3D streamlines of cuttings velocity for varying diameter ratios and inner pipe rotation with water as carrier fluid at 1.524 m/s.

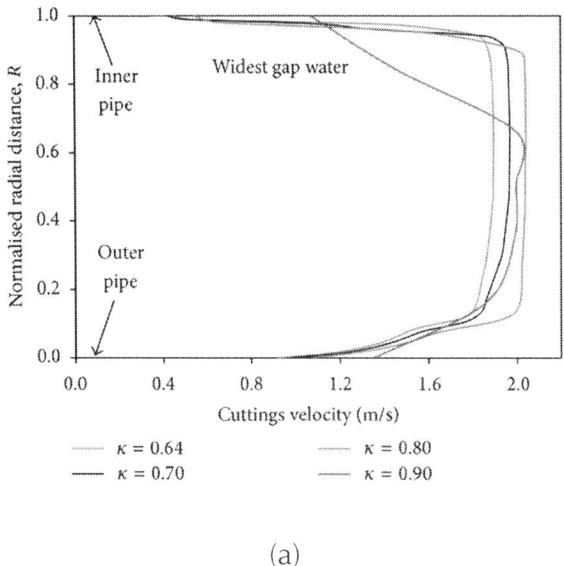

(a)

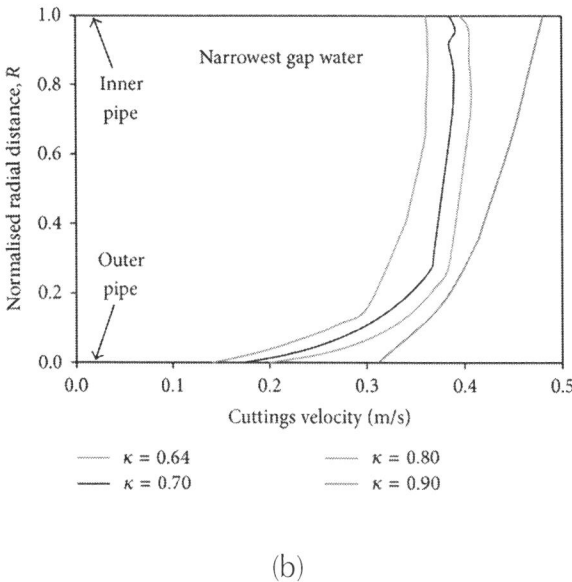

(b)

Figure 11: Cuttings velocity profiles with water as carrier fluid for varying diameter ratios at 1.524 m/s and 120 rpm: (a) gap above inner pipe and (b) gap below inner pipe.

Cuttings Volume Fraction, Velocity, and Profiles with Mud as Carrier Fluid

With mud as carrier fluid and flowing at 1.524m/s and a drill pipe rotating at 120 rpm, Figure 12 shows a very small cuttings volume fraction within the annular gap. Due to the high viscous nature of the mud, many cuttings are able to be suspended in the mud and then transported to the surface. This reduces the cuttings tendency to slip to the bottom of the wellbore to form a bed. The cuttings velocity presented in 3D streamlines (see Figure 13) shows how the cuttings travel in almost the entire annular space for all diameter ratios. This indicates better carrying capacity of the mud in transporting the cuttings to the surface. The radial measurements of the cuttings velocity profiles as shown in Figure 14 further illustrate the mud's carrying capacity in both the widest and narrowest annular gaps. The peak cuttings velocity also increases with increasing diameter ratio and is recorded in the widest gap as 1.698m/s, 1.758 m/s, 1.838m/s, and 1.840m/s for κ = 0.64, 0.70, 0.80, and 0.90, respectively, as shown in Figure 14(a). In the narrowest gap, as shown in Figure 14(b), the cuttings velocity profiles show irregular behaviours and are also very similar in magnitude for all diameter ratios. The peak cuttings velocities calculated are 1.000 m/s, 1.304m/s, 1.025 m/s, and 1.071 m/s for κ = 0.64, 0.70, 0.80, and 0.90, respectively (see Figure 14(b)).

CFD Method for Predicting Annular Pressure Losses... 31

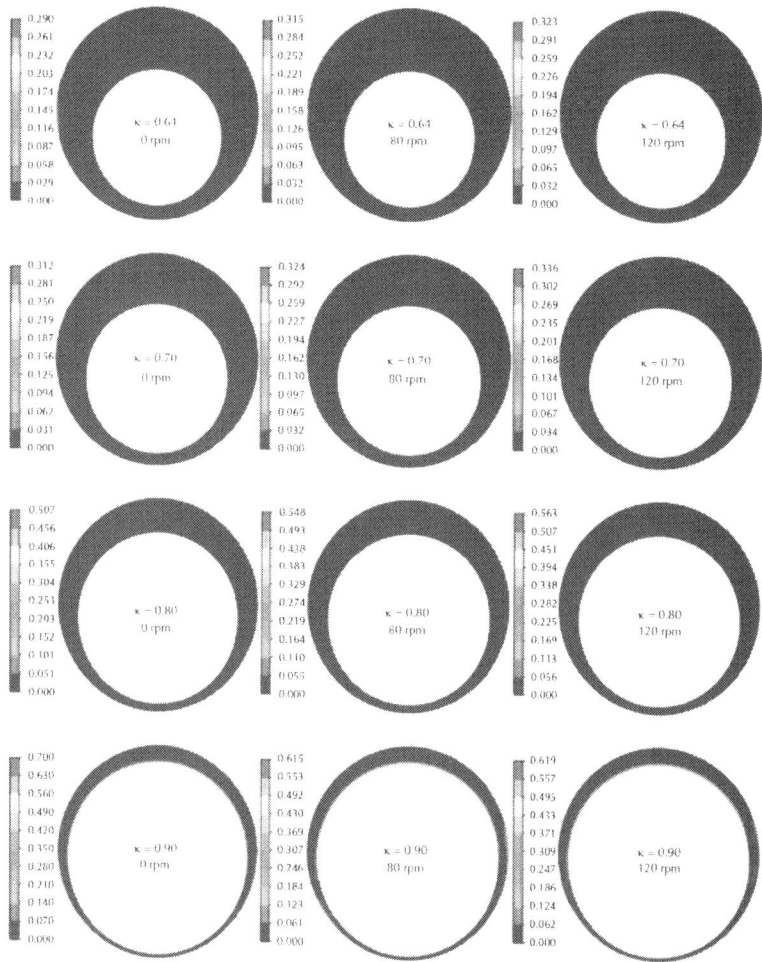

Figure 12: Contours of cuttings volume fraction for varying diameter ratios and inner pipe rotation with mud as carrier fluid at 1.524 m/s.

32   Drilling Engineering

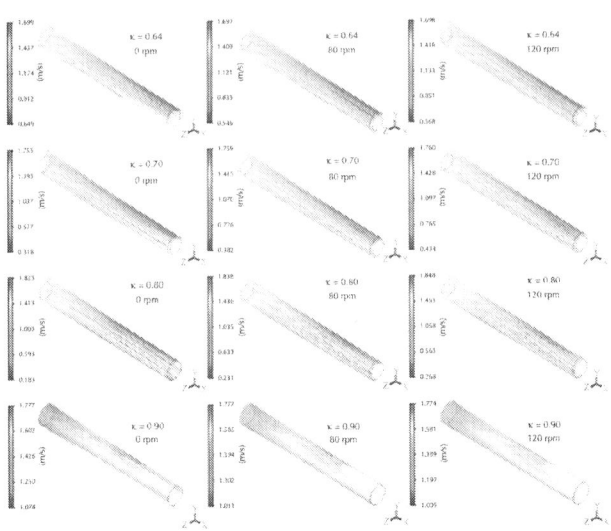

Figure 13: 3D streamlines of cuttings velocity for varying diameter ratios and inner pipe rotation with mud as carrier fluid at 1.524 m/s.

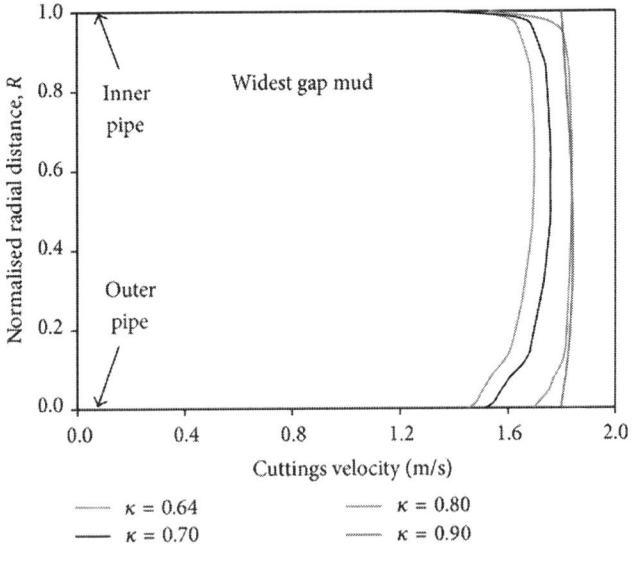

(a)

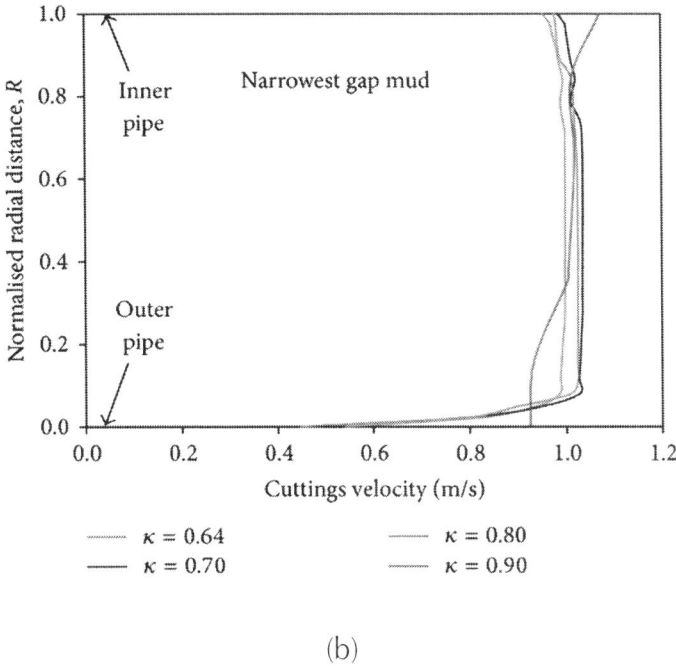

(b)

Figure 14: Cuttings velocity profiles with mud as carrier fluid for varying diameter ratio at 1.524 m/s and 120 rpm: (a) gap above inner pipe and (b) gap below inner pipe.

CONCLUSIONS

The present study employs a CFD method to analyse the effects of fluid velocity, annular diameter ratio (ranging from 0.64 to 0.90), drill pipe rotation, and fluid type on the prediction of pressure losses and cuttings concentration for solid-fluid flow in eccentric horizontal annular geometries. The following can be inferred from this study.

- Using water as carrier fluid, simulation data for pressure loss and cuttings concentration are in good agreement with experimental data with mean percentage errors of 0.84% and 12%, respectively. Similarly, with mud as carrier fluid, only

2.5% mean error exists between simulation and experimental pressure data, confirming the validity of the current model setup.

- Increasing annular fluid velocity significantly increases pressure losses, while a decrease in cuttings concentration occurs for each constant diameter ratio. This effect is however more pronounced for $\kappa = 0.90$ when using both water and mud as carrier fluids. Annular pressure loss is dramatically increased by 97% while cuttings concentration is decreased by 37% when the flowing mud velocity increased from 1.524m/s to 2.749m/s for $\kappa = 0.90$.
- When other drilling parameters are kept constant, increasing diameter ratio increases pressure loss, whereas a decrease in cuttings concentration is observed for each constant fluid velocity. This influence is however pronounced for $\kappa = 0.90$. Over 3600% increase in pressure loss could be realised while a decrease of about 86% in cuttings concentration is observed between diameter ratios of $\kappa = 0.64$ and $\kappa = 0.90$ for water flowing at a velocity of 1.524m/s.
- Increasing drill pipe rotation speed from 80 rpm to 120 rpm did not result in any significant increment in pressure losses with bothwater andmud. The rotation effect on cuttings concentration is quite predominant especially in annular gaps with diameter ratio below $\kappa = 0.70$ and at low fluid velocities. Contours of cuttings volume fraction show how rotation effect sweeps cuttings bed into the annularmainstreamand transports them to the surface.
- Although mud recorded higher pressure losses compared to water, it has better carrying capacity as opposed to water especially at smaller diameter ratios. The performance of both fluids on cuttings concentration is quite similar at high diameter ratios.

REFERENCES

1. P. H. Tomren, A. W. Iyoho, and J. J. Azar, "Experimental study of cuttings transport in directional wells," SPE Drilling Engineering, vol. 1, no. 1, pp. 43–56, 1986.
2. T. E. Becker and J. J. Azar, Mud-Weight and Hole-Geometry Effects on Cuttings Transport While Drilling Directionally, Society of Petroleum Engineers, SPE-14711-MS, 1985.
3. R. B. Adari, S. Miska, E. Kuru, P. Bern, and A. Saasen, "Selecting drilling fluid properties and flow rates for effective hole cleaning in high-angle and horizontal wells," in Proceedings of the SPE Annual Technical Conference and Exhibition, paper SPE-63050-MS, pp. 273–281, Dallas, Tex, USA, October 2000.
4. T. R. Sifferman and T. E. Becker, "Hole cleaning in full-scale inclined wellbores," SPE Drilling Engineering, vol. 7, no. 2, pp. 115–120, 1992.
5. R. Ahmed, M. Sagheer, N. Takach et al., "Experimental studies on the effect of mechanical cleaning devices on annular cuttings concentration and applications for optimizing ERD systems," inProceedings of the SPE Annual Technical Conference and Exhibition, paper SPE-134269-MS, pp. 2016–2028, Florence, Italy, September 2010.
6. M. E. Ozbayoglu, A. Saasen, M. Sorgun, and K. Svanes, "Critical fluid velocities for removing cuttings bed inside horizontal and deviated wells," Petroleum Science and Technology, vol. 28, no. 6, pp. 594–602, 2010.
7. J. O. Ogunrinde and A. Dosunmu, "Hydraulic optimization for efficient hole cleaning in deviated and horizontal wells," in Proceedings of the SPE Nigerian Annual Technical Conference and Exhibition, paper SPE 162970, Abuja, Nigeria, August 2012.
8. M. E. Ozbayoglu and M. Sorgun, "Frictional pressure loss estimation of water-based drilling fluids at horizontal and inclined drilling with pipe rotation and presence of cuttings,"

in Proceedings of the SPE Oil and Gas India Conference and Exhibition, paper SPE-127300-MS, Mumbai, India, January 2010.

9. M. Sorgun, I. Aydin, and M. E. Ozbayoglu, "Friction factors for hydraulic calculations considering presence of cuttings and pipe rotation in horizontal/highly-inclined wellbores," Journal of Petroleum Science and Engineering, vol. 78, no. 2, pp. 407–414, 2011.

10. O. M. Evren, E. Reza O, O. A. Murat, and Y. Ertan, "Estimation of "very-difficult-to-identify" data for hole cleaning, cuttings transport and pressure drop estimation in directional and horizontal drilling," in Proceedings of the IADC/SPE Asia Pacific Drilling Technology Conference and Exhibition, paper SPE-136304-MS, pp. 668–685, Ho Chi Minh City, Vietnam, November 2010.

11. N. C. G. Markatos, R. Sala, and D. R. Spalding, "Flow in an annulus of non-uniform gap," Transactions of the Institution of Chemical Engineers, vol. 56, no. 1, pp. 28–35, 1978.

12. S.-M. Han, Y.-K. Hwang, N.-S. Woo, and Y.-J. Kim, "Solid-liquid hydrodynamics in a slim hole drilling annulus," Journal of Petroleum Science and Engineering, vol. 70, no. 3-4, pp. 308–319, 2010.

13. M. Mokhtari, M. Ermila, A. N. Tutuncu, and M. Karimi, "Computational modelling of drilling fluids dynamics in casing drilling," in Proceedings of the SPE Eastern Regional Meeting, paper SPE-161301-MS, Lexington, Ky, USA, October 2012.

14. T. N. Ofei, S. Irawan, and W. Pao, "Modelling of pressure drop in eccentric narrow horizontal annuli with the presence of cuttings and rotating drillpipe," International Journal of Oil, Gas and Coal Technology. In press.

15. G. M. Faeth, "Mixing, transport and combustion in sprays," Progress in Energy and Combustion Science, vol. 13, no. 4, pp. 293–345, 1987.

16. M. Eesa and M. Barigou, "Horizontal laminar flow of coarse nearly-neutrally buoyant particles in non-Newtonian conveying fluids: CFD and PEPT experiments compared," International Journal of Multiphase Flow, vol. 34, no. 11, pp. 997–1007, 2008.
17. B. G. M. van Wachem and A. E. Almstedt, "Methods for multiphase computational fluid dynamics,"Chemical Engineering Journal, vol. 96, no. 1–3, pp. 81–98, 2003.
18. C. Y. Wen and Y. H. Yu, "Mechanics of fluidization," Chemical Engineering Progress Symposium Series, vol. 62, pp. 100–111, 1966.
19. D. Gidaspow, Multiphase Flow and Fluidization, Academic Press, 1994.
20. P. G. Saffman, "The lift on a small sphere in a slow shear flow," Journal of Fluid Mechanics, vol. 22, no. 2, pp. 385–400, 1965.
21. P. G. Saffman, "The lift on a small sphere in a slow shear flow—corrigendum," Journal of Fluid Mechanics, vol. 31, no. 3, p. 624, 1968.
22. R. Mei and J. F. Klausner, "Shear lift force on spherical bubbles," International Journal of Heat and Fluid Flow, vol. 15, no. 1, pp. 62–65, 1994.
23. B. E. Launder and D. B. Spalding, "The numerical computation of turbulent flows," Computer Methods in Applied Mechanics and Engineering, vol. 3, no. 2, pp. 269–289, 1974.
24. C. A. Shook and M. C. Roco, Slurry Flow: Principles and Practice, Butterworth-Heimemann, London, UK, 1991.
25. R. E. Osgouei, Determination of cuttings transport properties of gasified drilling fluids [Ph.D. thesis], Middle East Technical University, Ankara, Turkey, 2010.
26. S. V. Patankar, Numerical Heat Transfer and Fluid Flow, Hemisphere Publishing Corp., 1980.

Decorating and Filling of Multi-walled Carbon Nanotubes with TiO2 Nanoparticles via Wet Chemical Method

Sedigheh Abbasi[1], Seyed Mojtaba Zebarjad[2], and Seyed Hossein Noie Baghban[1]

[1]Department of Chemical Engineering, Faculty of Engineering, Ferdowsi University of Mashhad, Mashhad, Iran

[2]Department of Material Science and Engineering, Faculty of Engineering, Ferdowsi University of Mashhad, Mashhad, Iran

ABSTRACT

Multi-walled carbon nanotubes (MWCNTs) have been successfully modified with TiO_2 nanoparticles via wet chemical method. For this purpose tetra chloride titanium ($TiCl_4$) was used as titanium source.

MWCNTs were exposed at different amount of $TiCl_4$ (0.25 and 0.1 ml) and different soaking times. The modified MWCNTs have been characterized by X-ray diffraction (XRD) and transmission electron microscopy (TEM). TEM results showed that the MWCNTs were fully decorated with TiO_2 at short term immersion. Increasing soaking time caused to fill the MWCNTs with TiO_2 nanoparticles. The results showed that the amount of precursor had a significant role on quantity of decoration. The decoration of outer surface of MWCNTs with TiO_2 was more noticeable at large amount of $TiCl_4$. XRD results revealed that the crystalline structure of TiO_2 on the surface and inner of MWCNTs was rutile. The average size of TiO_2 nanoparticles which modified MWCNTs were 20 nm.

INTRODUCTION

In the recent years, various publications have reported the outstanding physicochemical characteristics of Carbon Nanotubes (CNTs) [1]. The study of these nanostructures lead to diagnose individual properties such as high strength to weight ratio, excellent Young's modulus which has been measured to reach about 1 TPa, exceptional mechanical and electrical properties and high flexibility [2]. The mentioned properties cause to appear many applications in different field of researches and new technologies [3]. For instance excellent elasticity and flexibility of CNTs plays an important role for choosing CNTs as alternative reinforcing filler in composite materials. Numerous authors have reported the incorporation of these tubular structures in diverse matrices, although the most reproducible and successful advances have been achieved when fabricating ceramic and polymer composites [4-7].

Modification of CNTs has been paid a great deal of interest as a fascinating model system for fundamental scientific research with the potential technological applications. In particular, fabrication of CNT-based devices which modified with metal oxides is more applicable than non-modified devices. Up to now, various metal oxides such as TiO_2, SnO_2, ZnO, Fe_2O_3, MnO_2 and RuO_2 have been reported to modify CNTs [8-13]. Nowadays finding a simple and low-cost approach to modify CNTs with the metal oxides is still an active research. Most published work to date has focused on the coating and filling of carbon nanotubes with metals and metal oxides. However, the difficulty of obtaining a uniform coated nanotube imposes some limitations. Multi-walled carbon nanotube (MWNT) based metal oxide composites [14] were prepared by an impregnation method using organometallic compounds as precursor. Tasviri et al. [15] obtained TiO_2-coated carbon nanotubes which were functionalized with amine groups. Bouazza et al. [6] have coated Multi-walled carbon nanotubes with TiO_2 layer by sol-gel method. The CNTs-TiO_2 composites fabricated by hydrolysis in which the formation of TiO_2 and its compounding with CNTs happened almost simultaneously [16] have also been reported. Chun Oh et al. [7] synthesized CNT/TiO_2 composites by an improved oxidation method.

CNTs and titanium dioxide (TiO_2) composite materials have attracted attention of researchers in relation to the treatment of contaminated water and air by heterogeneous photo-catalysis, hydrogen evolution, CO_2 photoreduction, and dye sensitized solar cells. These composite materials have been fabricated by a range of different methods, including mechanical mixing of TiO_2 and CNTs [17], sol-gel synthesis of TiO_2 in the presence of CNTs [18], electro-spinning methods [19], electrophoretic deposition [20], and chemical vapor deposition [8]. The uniformity of the decoration of CNTs with oxide and the physical properties of the composite materials depend on the procedure of modification. For example uniform decorating of CNTs with TiO_2 nanoparticles have been reported by chemical vapor deposition and electro-spinning methods [21]. However these methods are complex and need special equipments. The modified CNTs with sol-gel methods lead to heterogeneous and nonuniform decorations and aggregation of TiO_2 on the surface of CNTs [22].

Here we produced modified CNTs with TiO_2 by a new and simple two-step wet chemical method. The preparation of modified CNTs by this technique is very simple and doesn't need to special equipments.

MATERIALS AND EXPERIMENTAL

For preparation of the synthesized CNT-TiO$_2$ the following materials were used:

Multi-walled carbon nanotubes (MWCNTs, 95.9% purity, diameter: ~40 - 60 nm, length: ~5 - 15 μm), Tetra chloride titanium (TiCl$_4$, M = 189.79, 99%, Merck), Nitric Acid (HNO$_3$, M = 63, 65%, Merck), Hydrochloride Acid (HCl 37 wt%, Merck).

A typical experimental procedure was followed. Firstly, the MWNTs were opened by oxidation with nitric acid solution (65%) at room temperature for 2 h in an ultrasound bath and then the mixture was stirred at high speed for 2 h, and the treated MWCNTs were washed with distilled water several times until the pH reached 7 and dried at 90°C for an overnight. Secondly a certain amount of TiCl$_4$ was added to 100 ml of distilled water, followed by adding a little HCl (37 wt%) to the distilled water before TiCl$_4$ was dissolved in the water. The addition of a small amount of acid is only for reducing the hydrolysis of TiCl$_4$. 75 mg of acid treated MWCNTs were dispersed in this solution using ultrasound bath for 2 h. The mixture was stirred for 22 h at room temperature and then raised the temperature up to 80°C. The mixture was stirred for 3 h then filtered and dried at 80°C for 1 h and calcinated at 370°C for 3 h. A flow chart of the technique is shown in Figure 1. X-ray powder diffraction was performed to characterize the phase composition and crystal structure of the samples, using a Philips Analytical X-Ray B.V equipment (40 kV/30 mA) with Cu Kα (1.542 Å) radiation. The scanning velocity was 0.02° s^{-1}, and the 2θ range scanned ranged from 4° to 90°.

The crystalline sizes of the produced phase were determined by the Scherrer formula, using a K factor of 0.9:

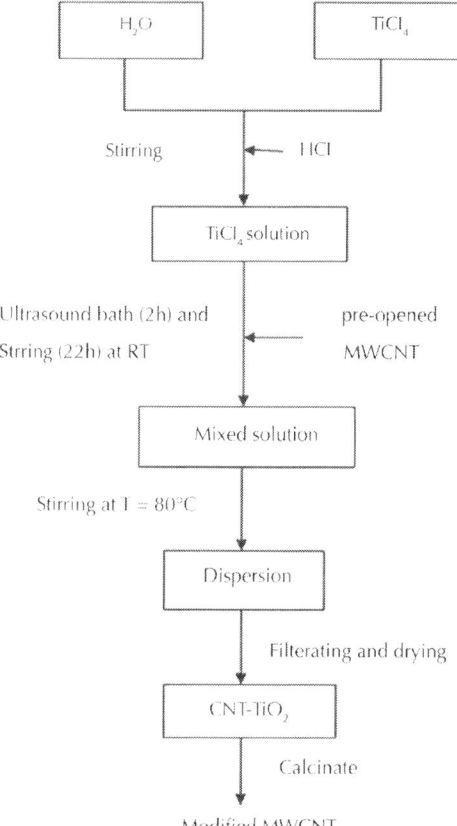

Figure 1: Flow chart of the steps involved in the preparation modified MWCNT.

$$B = \frac{KL}{\beta \cos \theta} \tag{1}$$

where B: crystalline size, in nm; L: wavelength for the radiation used, which is 1.5406°A for Cu; β: full width at half maximum intensity (FWHM); θ: angle for the XRD maximum peak.

The weight fractions of the TiO_2 nanoparticle and CNT in the modified MWCNT were calculated from the relative intensities of the strongest peaks corresponding to TiO_2 and graphite as described by Spurr and Myers [23]:

$$X = \frac{1}{\left[1+0.8\left(I_R/I_G\right)\right]} \qquad (2)$$

where X is the weight fraction of CNT in the product, while I_R and I_G are the X-ray integrated intensities of the (110) reflection of rutile and (002) reflection of graphite, respectively.

The transmission electron microscopy (TEM, LEO 912 AB.) was used to investigation the quality of modification and measurement of inner and outer diameter of CNTs and particle size of formed TiO_2.

RESULTS AND DISCUSSION

Oxidation of MWCNTs

Figure 2 illustrates the opening of MWCNTs by oxidative phenomenon. The TEM micrographs show opening of the MWCNTs at the end tips at different magnifications. It can be seen that the ends of some tubes were opened and would cut to short length. This is because the hexagon electrophilic destroyed by acid on the integrated graphene structures probably cut the nanotubes down to short tubes [24].

Modification of MWCNTs

Figure 3 shows one of the low-magnification TEM images of modified MWCNT, which reveals that the outer surface of MWCNTs is fully coated with TiO_2 nanocrystals.

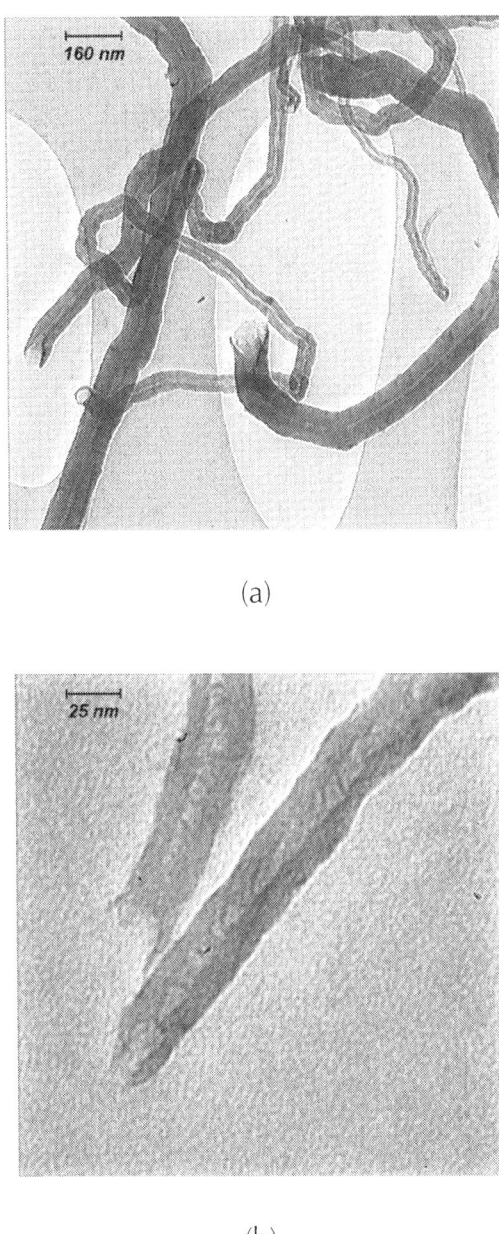

(a)

(b)

Figure 2: TEM images of opened multiwall carbon nanotube at different scales.

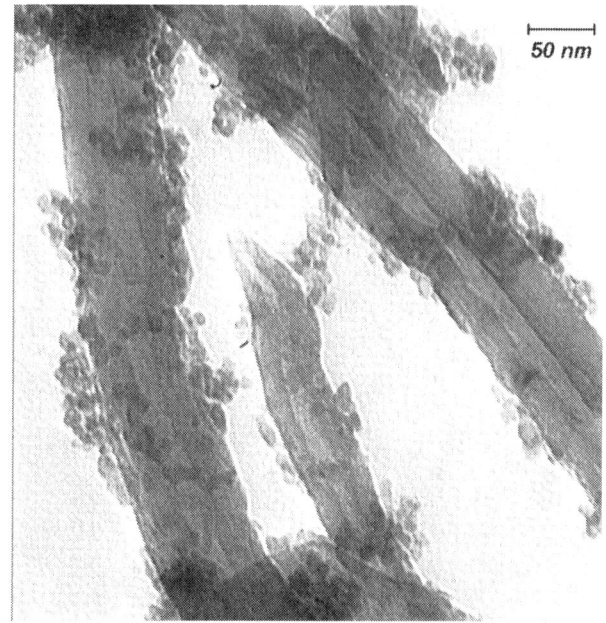

Figure 3: TEM image of outer surface decoration of MWNTs.

The method which we used for modification of MWCNTs involves three steps. Firstly, negatively charged functionalities such as -COOH and -OH are introduced to the surface of MWCNTs after oxidation [25]. Then, the titanium ions in the solution which produced by hydrolyses of $TiCl_4$ are adsorbed to the surfaces due to the electrostatic attraction. Finally, TiO_2 nanocrystals are in-situ formed on the outer surface of MWCNTs. The mechanism of nano size formation of TiO_2 is as below:

$$TiCl_4 + 2H_2O \rightarrow TiO_2 + 4HCl \tag{3}$$

Effect of Time on Modification

The result of experiments revealed that soaking time has an effective influence on modification of samples. The dominant mechanism in short-term immersion (24 h) is decoration of outer surface with TiO_2. At the same time the inner of MWCNTs are filled with TiO_2

nanoparticles. They are most noticeable because of the striking contrast from the filling material to the empty cavity of MWCNTs, which is consistent with the presence of TiO_2 nanoparticles (Figure 4). It can be clearly seen that the fine TiO_2 nanoparticles attach to one another densely and continuously, and form an almost continuous dark line in the cavity of the MWCNTs. The average size of the TiO_2 nanoparticle is about 20 nm.

Figure 5 shows the filling of MWCNTs with TiO_2 due to its exposure under $TiCl_4$ solution for 48 h. As seen there is obvious filled materials to the empty cavity of MWCNTs. With comparison of Figure 4 and Figure 1 may conclude that the amount of filled TiO_2 depends strongly on the soaking time in $TiCl_4$ solution.

Figure 4: TEM image of decoration of outer surface and filling of MWNTs at short mixing duration (24 h).

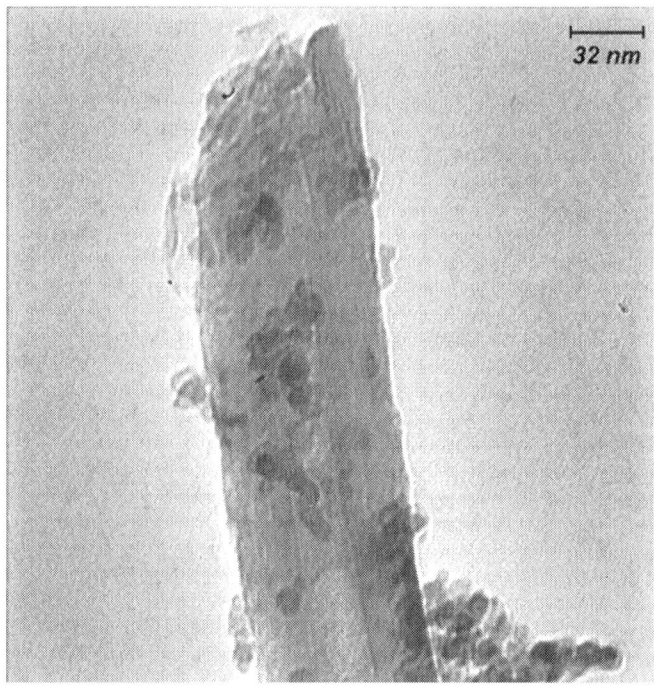

Figure 5: TEM image of decoration of outer surface and filling of MWNTs at long mixing duration (48 h).

Effect of TiCl$_4$ Content on Modification

As mentioned before the amount of TiCl$_4$ was added to 100 ml of distilled water at the same amount of HCl and MWCNT is an important factor for the decoration of MWCNTs with TiO$_2$. For instance by increasing the amount of TiCl$_4$, the probability of decoration of outer surface of MWCNTs will be increased. As shown in Figures 6(a) and (b) at large amount of TiCl$_4$ decoration is more obvious. It can be concluded that at large amount of TiCl$_4$, the titanium ions in the solution are more and further titanium ions are adsorbed to the surfaces due to the electrostatic attraction and hence decoration of outer surface of MWCNTs with TiO$_2$ nanoparticles is large.

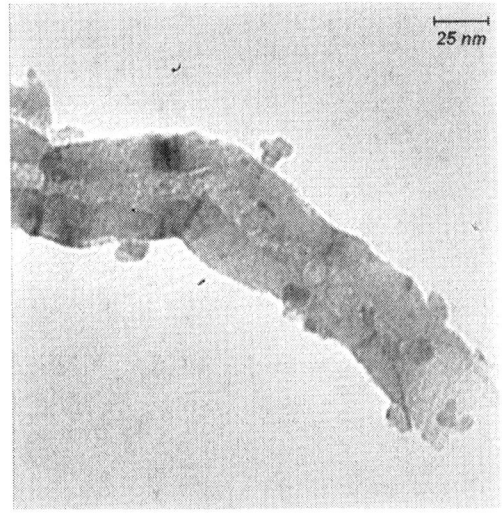

(a)

(b)

Figure 6: TEM micrographs of decoration of MWCNTs at different amount of TiCl$_4$, (a) A little amount; (b) A large amount.

Characterization

Figure 7 shows the XRD pattern of the pristine MWCNTs (a) and MWCNTs-TiO$_2$ (b) which reveals various crystal structures. The XRD pattern of MWCNT exhibits a sharp peak at around 2θ = 26° and a broad peak centered at 2θ = 43° corresponding to the (002) and (100) Bragg reflection planes having interlayer spacing of 3.348 and 2.027 A respectively. The observed diffractions in sample MWCNTs-TiO$_2$ are displayed both the peaks assigned to the MWCNT and TiO$_2$, respectively, and their broadness suggests the presence of very small crystals. The peak at a 2θ of 26.603° is typical for the (002) diffraction of graphite [6] and confirms the presence of CNTs in the samples.

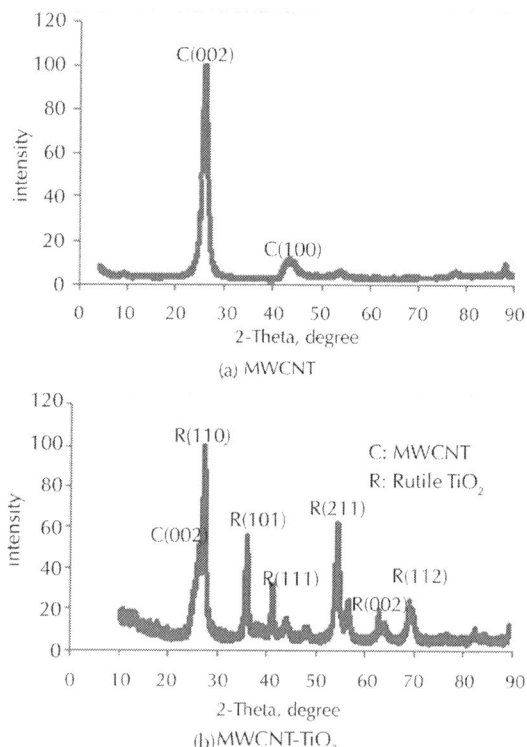

Figure 7: XRD patterns of (a) Pristine MWCNTs and (b) CNT-TiO$_2$.

Notice that there are no traces of anatase in this sample. The peaks at 27.387°, 36.007°, 41.158°, and 54.213° belong to the diffraction peaks of (110), (101), (111), and (211) of rutile. Therefore, it can be concluded that the CNT/TiO$_2$ had a structure like rutile crystals. As XRD analysis confirmed, the average size of the TiO$_2$ crystallite is about 13.21 nm in accordance with the results calculated using the Scherrer equation which mentioned in Equation (1). Modified MWCNT consists of both TiO$_2$ nanoparticle and graphite at the ratio X = 0.401 using the Spurr and Myers equation that means the weight fraction of MWCNT in the sample is 0.401.

CONCLUSIONS

In summary, we have demonstrated a simple and reproducible wet chemical method to modify MWCNTs with TiO$_2$ nanoparticles using TiCl$_4$ as precursor. The amount of TiCl$_4$ and soaking time play critical roles in modification of MWCNTs. According to the TEM results, the average size of the TiO$_2$ nanoparticle is about 20 nm. In the XRD pattern, the sample was observed to be a phase of rutile and their broadness suggests the presence of very small crystals. It is worth noting that by using this method one can obtain different amount of TiO$_2$ nanocrystals modified MWCNTs only by changing the concentration of precursor.

REFERENCES

1. M. Terrones, "Carbon Nanotubes: Synthesis and Properties, Electronic Devices and Other Emerging Applications," International Materials Reviews, Vol. 49, No. 6, 2004, pp. 325-377.
2. C. Li and T. Chou, "Elastic Moduli of Multi-Walled Carbon Nanotubes and the Effect of van der Waals Forces," Composites Science and Technology, Vol. 63, No. 11, 2003, pp. 1517-1524. doi:10.1016/S0266-3538(03)00072-1

3. M. S. Dreselhaus, G. Dresselhaus and P. Avouris, "Carbon Nanotubes: Synthesis, Structure, Properties and Applications," Springer, Berlin, 2001. doi:10.1007/3-540-39947-X
4. E. T. Thonstenson, Z. Ren and T. W Chou, "Advances in the Science and Technology of Carbon Nanotubes and Their Composites: A Review," Composites Science and Technology, Vol. 61, No. 13, 2001, pp. 1899-1912. doi:10.1016/S0266-3538(01)00094-X
5. W. A. Curtin and B. W. Sheldon, "Review: Ceramic and Metal Nanocomposites," Materials Today, Vol. 7, No. 11, 2004, pp. 44-48. doi:10.1016/S1369-7021(04)00508-5
6. N. Bouazza, M. Ouzzine, M. A. L. denas, D. Eder and A. L. Solano, "TiO_2 Nanotubes and CNT-TiO_2 Hybrid Materials for the Photocatalytic Oxidation of Propene at Low Concentration," Applied Catalysis B: Environmental, Vol. 92, No. 3-4, 2009, pp. 377-383.doi:10.1016/j.apcatb.2009.08.017
7. W. Chun Oh, M. L. Chen and B. K. Chem, "Synthesis and Characterization of CNT/TiO_2 Composites Thermally Derived from MWCNT and Titanium(IV) n-Butoxide," Bulletin of the Korean Chemical Society, Vol. 29, No. 1, 2008.
8. C. S. Kuo, Y. H. Tseng, H. Y. Lin, C. H. Huang, C. Y. Shen, Y. Y. Li, S. I. Shah and C. P. Huang, "Synthesis of a CNT-Grafted TiO(2) Nanocatalyst and Its Activity Triggered by a DC Voltage," Nanotechnology, Vol. 18, No. 46, 2007, Article ID: 465607.doi:10.1088/0957-4484/18/46/465607
9. G. M. An, N. Na, X. R. Zhang, Z. J. Miao, S. D. Miao, K. L. Ding and Z. M. Liu, "SnO2/Carbon Nanotube Nanocomposites Synthesized in Supercritical Fluids: Highly Efficient Materials for Use as a Chemical Sensor and as the Anode of a Lithium-Ion Battery," Nanotechnology, Vol. 18, No. 43, 2007, Article ID: 435707. doi:10.1088/0957-4484/18/43/435707
10. X. Y. Wang, B. Y. Xia, X. F. Zhu, J. S. Chen, S. L. Qiu and J. X. Li, "Controlled Modification of Multiwalled Carbon Nanotubes with ZnO Nanostructures," Solid State Chemistry, Vol. 181, No. 4, 2008, pp. 822-827. doi:10.1016/j.jssc.2008.01.005

11. J. W. Liu, X. J. Li and L. M. Dai, "Water-Assisted Growth of Aligned Carbon Nanotube-ZnO Heterojunction Arrays," Advanced Materials, Vol. 18, No. 3, 2006, pp. 1740-1744. doi:10.1002/adma.200502346
12. X. B. Fan, F. Y. Tan, G. L. Zhang and F. B. Zhang, "A Novel Strategy to Fabricate γ-Fe_2O_3-MWCNTs Hybrids with Selectively Ferromagnetic or Superparamagnetic Properties," Materials Science and Engineering A, Vol. 454-455, 2007, pp. 37-42.doi:10.1016/j.msea.2007.01.027
13. Z. Y. Wang, G. Chen and D. G. Xia, "Coating of MultiWalled Carbon Nanotube with SnO_2 Films of controlled Thickness and Its Application for Li-Ion Battery," Power Sources, Vol. 184, No. 2, 2008, pp. 432-436. doi:10.1016/j.jpowsour.2008.03.028
14. K. Hernadi, E. Ljubovic, J. W. Seo and L. Forro, "Synthesis of MWNT-Based Composite Materials with inorganic Coating," Acta Materialia, Vol. 51, No. 5, 2003, pp. 1447-1452. doi:10.1016/S1359-6454(02)00539-6
15. M. Tasviri, H. A. R. Pourb, H. Ghourchianb and M. R. Gholami, "Amine Functionalized TiO_2 Coated on Carbon Nanotube as a Nanomaterial for Direct Electrochemistry of Glucose Oxidase and Glucose Biosensing," Molecular Catalysis B: Enzymatic, Vol. 68, No. 2, 2011, pp. 206- 210. doi:10.1016/j.molcatb.2010.11.005
16. L. Chen, B. L. Zhang, M. Z. Qu and Z. L. Yu, "Preparation and Characterization of CNTs-TiO_2 Composites," Powder Technology, Vol. 154, No. 1, 2005, pp. 70-72.doi:10.1016/j.powtec.2005.04.028
17. C. Y. Kuo, "Prevenient Dye-Degradation Mechanisms Using UV/TiO_2/Carbon Nanotubes Process," Hazardous Mater, Vol. 163, No. 1, 2009, pp. 239-244.doi:10.1016/j.jhazmat.2008.06.083
18. W. Wang, P. Serp and P. Kalck, "Photocatalytic Degradation of Phenol on MWNT and Titania Composite Catalysts Prepared by a Modified Sol-Gel Method," Applied Catalysis B: Environmental, Vol. 56, No. 4, 2005, pp. 305-312. doi:10.1016/j.apcatb.2004.09.018

19. S. Aryal, C. K. Kim and K. W. Kim, "Multi-Walled Carbon Nano-Tubes/TiO2 Composite Nanofiber by Electrospinning," Materials Science and Engineering: C, Vol. 28, No. 1, 2008, pp. 75-79. doi:10.1016/j.msec.2007.10.002
20. J. Cho, S. Schaab and J. A. Roether, "Nanostructured Carbon Nanotube/TiO_2 Composite Coatings Using Electrophoretic Deposition (EPD)," Journal of Nanoparticle Research, Vol. 10, No. 1, 2008, pp. 99-105. doi:10.1007/s11051-007-9230-x
21. H. Yu, X. Quan and S. Chen, "Aligned TiO_2-Multiwalled Carbon Nanotube Heterojunction Arrays and Their Charge Separation Capability," The Journal of Physical Chemistry C, Vol. 111, No. 35, 2007, pp. 12987-12991.
22. J. Sun, L. Gao and M. Iwasa, "Noncovalent Attachment of Oxide Nanoparticles onto Carbon Nanotubes Using Water-in-Oil Microemulsions," Chemical Communications, Vol. 7, 2004, pp. 832-833. doi:10.1039/b400817k
23. Q. H. Zhang, L. Gao and J. K. Guo, "Preparation and Characterization of Nanosized TiO_2 Powders from Aqueous $TiCl_4$ Solution," Nanostructured Materials, Vol. 11, No. 8, 1999, pp. 1293-1300. doi:10.1016/S0965-9773(99)00421-3
24. J. Zhang, H. Zou, Q. Qing, Y. Yang, Q. Li, Z. Liu, X. Gou and Z. Du, "Effect of chEmical Oxidation on the Structure of Single-Walled Carbon Nanotubes," The Journal of Physical Chemistry B, Vol. 107, No. 16, 2003, pp. 3712-3718. doi:10.1021/jp027500u
25. L. Zhao and L. Gao, "Filling of Multi-Walled Carbon Nanotubes with tin(IV) Oxide," Carbon, Vol. 42, No. 15, 2004, pp. 3251-3272.

Study of Sodium-Chromium-Iron-Phosphate Glass by XRD, IR, Chemical Durability and SEM

Youssef Makhkhas[1], Said Aqdim[2], and El Hassan Sayouty[1]

[1]Hassan II University of Casablanca, Faculty of Science, High Energy and Condensed Matter Lab, Casablanca, Morocco

[2]Hassan II University of Casablanca, Faculty of Science, Mineral chemistry Lab, Casablanca, Morocco

ABSTRACT

Chromium iron phosphate glass was investigated for use as waste form because of its improved chemical durability. The introduction of chromium in sodium-iron-phosphate glass is used to compare its effect with iron in inhibition of corrosion. The sodium-chromium-iron phosphate glass of composition $10Na_2O-30Fe_2O_3-5Cr_2O_3-55P_2O_5$ (mol %) was produced by melting batches of (99, 98% pure) Cr_2O_3, Fe_2O_3, Na_2CO_3, and $(NH_4)_2HPO_4$ at 1080°C for one hour and pouring the liquid into steel mold. The sample was annealed at 680°C for 48 h. We have performed the measurement of X-Ray Diffraction (XRD), Scanning Electronic Microscopy (SEM), Infra-Red spectroscopy (IR), and the chemical durability. The IR of the glass studied, contains two dominant bands, which were characteristic of pyrophosphate groups, (P-O) stretching mode of P-O non bridging oxygen at 1055 cm^{-1} and sym stretching mode of bridging oxygen at 444 cm^{-1} respectively. There is also a band at 603 cm^{-1} attributed to isolated tetrahedral units $(PO_4)^{3-}$. The chemical durability of the glass was investigated by measuring the weight loss in distilled water at 90°C for 22 days.

INTRODUCTION

Phosphate glasses are of particular interest in both technological and scientific fields because they generally have lower processing temperatures less than 1000°C, lower glass transition temperatures (Tg) [1-6], and higher thermal expansion coefficients (α) than silicate glasses in the range of 90 to 250 × 10^{-7}/°C [4-6]. These properties makes them a good candidates in many applications such as glass to metal seals, thick film paste, the molding of optical elements, low temperature enamels for metals [2-4]. However, their relatively poor chemical durability makes them unsuitable for practical applications [7-9]. It was reported [9-15] that the introduction of oxides such as SnO, PbO, ZnO, Cr_2O_3 and Fe_2O_3, results in the formation of Sn-O-P, Pb-O-P, Zn-O-P, P-O-Cr and P-O-Fe bonds, and leads to improvement in the chemical durability of the modified phosphate glasses.

The iron phosphate glasses have generally both excellent chemical durability and low melting temperature typical between 950 and 1100°C, [16]. Chromium phosphate glasses for the immobilisation and disposal of nuclear waste were reported in 1984

[17]. The combination of chromium phosphate glass with various types of simulated nuclear waste showed that it is possible to have a waste form with a corrosion rate more slowly than that one of a comparable borosilicate glass. Therefore it has been suggested that the chemical durability of sodiumchromium-iron phosphate glasses is attributed to the replacement of P-O-P bonds by P-O-Cr and P-O-Fe bonds. The presence of P-O-Fe bands in higher concentrations, makes the glass more hydration resistant [16,18-21]. The P-O-Cr bands seem to play the same role than P-O-Fe bands [22].

In this paper we present a study of sodium chromium iron phosphate glass $10Na_2O-30Fe_2O_3-5Cr_2O_3-55P_2O_5$ which is prepared by the melt quenching technique, and characterized by X-Ray Diffraction, Infra-Red spectroscopy, and Scanning Electronic Microscopy. The chemical durability was investigated in distilled water solution.

EXPERIMENTAL

The glass of composition $10Na_2O-30Fe_2O_3-5Cr_2O_3-55 P_2O_5$ (mol %) is obtained by the melting quench method in 1080°C. Appropriate mixture of mixing compounds Na_2CO_3, ferric oxides, Cr_2O_3 and $(NH_4)_2HPO_4$ were initially tempered at various temperatures between 300°C - 500°C to achieve a preparation before the glass preparation.

The melt was achieved in alumina crucibles for about 30 mn at 1080°C ± 10°C. The isolated glasses samples have an approximate of size 10 mm diameter and 3 mm in thickness. The vitreous state was first evidenced from the shiny aspect and then confirmed from XRD patterns. Annealing of this glass was realized at 680°C for 48 hours. The first structural approach was made using X-rays diffraction which allowed to following. The density of the glass was measured at room temperature using the helium pycnometry method. The chemical durability sample of the size 0.9 × 0.9 × 0.9 × 0.3 cm was used sample were first polished to a 400 grit finished with SiC paper, then they were immersed in a flask filled with 100 ml of distilled water at 90°C for a time of 22 days. The dissolution rate (D_R) was then determined from the weight loss during the aqueous treatment at 90°C. The infrared (IR) spectra for each glass were measured between 400 and 1600 cm^{-1} using mX^{-1} and NIC-3600 FTIR spectrometers. The sample was prepared by pressing a mixture of about 2 mg of glass powder and 100 mg of anhydrous KBr powder. The chemical of composition of the analysed glass is given in Table 1.

X-Ray Diffraction, and Density

No crystalline phase was detected by X-ray in the glass composition 10Na$_2$O-30Fe$_2$O$_3$-5Cr$_2$O$_3$-55P$_2$O$_5$, whereas we notice that there is the existence of crystalline micro domain which means that the crystallization of the glass had just started, this observation comes from apearence of some peaks related to the crystalline phase NaFeP$_2$O$_7$ (Figure 1(a)). The XRD pattern of annealed glass (Figure 1(b)) shows more crystallization peaks of the same phase, this result might confirm that the glass structure could be composed by NaFeP$_2$O$_7$ pyrophosphates groups.

Infrared Spectroscopy Study

The infrared spectra shown in Figure 2, for the glass of composition 10Na$_2$O-30Fe$_2$O$_3$-5Cr$_2$O$_3$-55P$_2$O$_5$ include two dominant bands. These bands are characteristics of pyrophosphate groups: the band at 1099 and 760 cm^{-1} are attributed to (PO$_3$)$^{1-}$ asymmetric stretching mode of non-bridging oxygen (P-O-P) sym stretching mode of bridging oxygen, respectively [22-25]. The band at 992 cm^{-1} seems to be assigned to the isolated tetrahedral units (PO$_4$)$^{3-}$ [26-28].

SEM Microscopy Study

We noticed that the SEM plots (Figures 3(a) and (b)) for both glasses before and after aqueous dissolution confirms the presence of all the elements forming the glass. The proportions of the constituents seem as well to be in accordance with the theoretical composition.

The microscopic photograph was made at the nanometrical scale (as shown in Figures 4(a) and (b)).

The surface of the glass before the aqueous attack appears to be homogeneous in comparison with the attacked one which shows a deteriorated zone on its surface. Taking in consideration the dimension of the photographs, the etched surface remains very small.

Table 1: bench composition, calculated (O/P) ratio, D_R and density of the glass studied

Starting glass composition (mol%)				[O/P] ratio	(DR) (g/cm² / mn)	r (g/cm³)
Na_2O	Fe_2O_3	Cr_2O_3	P_2O_5	22 days		±0.02
10	30	5	55	3.72	$(7.188 \pm 0.001) \times 10^{-9}$	2.627

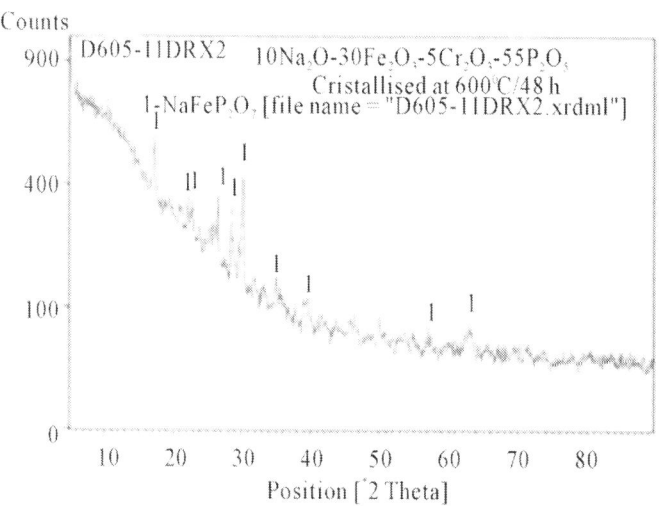

(a)

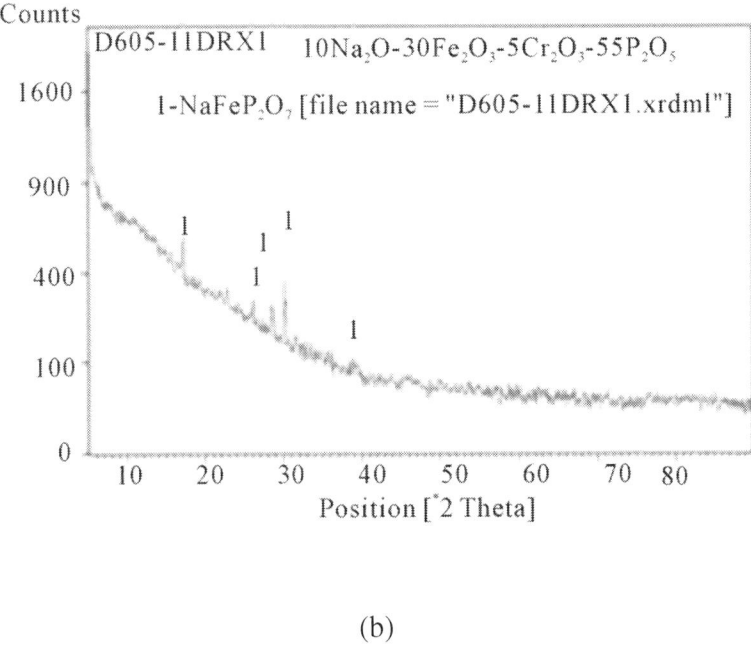

(b)

Figure 1: (a) X-ray diffraction spectra for the sample $10Na_2O$-$30Fe_2O_3$-$5Cr_2O_3$-$55P_2O_5$; (b) X-Ray diffraction pattern for $10Na_2O$-$30Fe_2O_3$-$5Cr_2O_3$-$55P_2O_5$ glass partially crystallized at 650°C for 48 h.

DISSOLUTION RATE

The dissolution rate (D_R) for sodium-lead-iron-phosphate glass is a function of time (see Figure 5). The dotted line indicates the dissolution rate of the glass which is included for comparison. Experimental dissolution rate for lead-iron-phosphate glass were reported by Day and et al. [29]. In the present work the dissolution rate for $10Na_2O$- $30Fe_2O_3$-$5Cr_2O_3$-$55P_2O_5$ glass decreases as the time of immersion in water increases. This decrease is observed in this type of glass because in the early stage the solution is still dilute, and the increase of glass leaching product has a relatively minor effect upon the rate of reaction.

DISCUSSION-CORRELATION BETWEEN THE STRUCTURE AND THE HIGH DURABILITY OF IRON PHOSPHATE GLASSES

Both XRD and IR techniques have confirmed the structural evolution of the glass network towards the pyrophosphate. So the structure of sodium-chromium-iron phosphate glass can be considered as pyrophosphate units connected with ferric and ferrous ions in octahedral or distorted octahedral coordination [30]. The chemical durability of sodium-chromium-iron phosphate glass of composition $10Na_2O-30Fe_2O_3-5Cr_2O_3-55P_2O_5$ (regarding aqueous attack at 90°C) is attributed to the increasing number of Fe-O-P bonds in the glass [30,31]. Such bonds are expected to be more water resistant than the P-O-P and Na-O-P bands [32]. This glass have a (D_R) 50 times less than the D_R of window glass and ~150 times less than the D_R for BABAL glass which have been considered as alternative materials for the immobilization of nuclear waste substance [32,33].

CONCLUSIONS

The structure and the chemical durability of sodiumchromium-iron phosphate glass of composition $10Na_2O-30Fe_2O_3-5Cr_2O_3-55P_2O_5$ (mol %) have been investigated using various techniques such IR, XRD, SEM.

The structure of the $10Na_2O-30Fe_2O_3-5Cr_2O_3-55P_2O_5$ (mol %) glass can be considered as pyrophosphate units connected with chromium. This glass have a (D_R) ~200 times less than the D_R for BABAL glass which have been considered as alternative materials for the immobilization of nuclear waste substance. This result is very important for applications in the nuclear waste management. The improved chemical durability is attributed to the replacement of the easily hydrated Na-O-P and P-O-P bonds by corrosion resistant Fe-O-P and Pb-O-P bands. The recorded IR spectra indicate that these glasses are dominated by $(P_2O_7)^{4-}$ dimmer units, and contain a large number of Fe-O-P bonds.

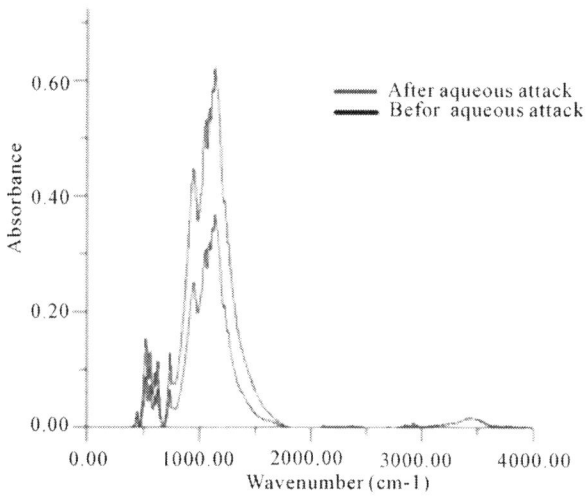

Figure 2: The IR spectra of sodium-chromium-iron-phosphate glass.

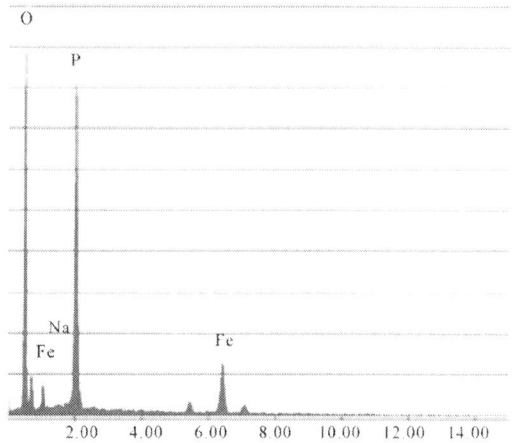

(a)

Study of Sodium-Chromium-Iron-Phosphate Glass.... 63

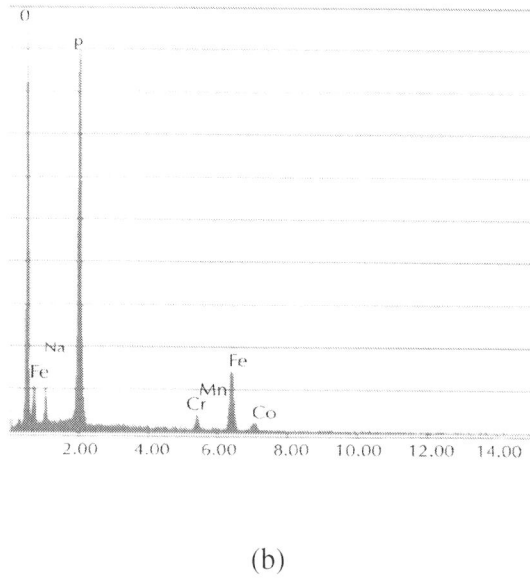

(b)

Figure 3: (a) The SEM plot for the sample glass $10Na_2O-30Fe_2O_3-5Cr_2O_3-55P_2O_5$ (before aqueous attack); (b) The SEM plot for the sample glass $10Na_2O-30Fe_2O_3-5Cr_2O_3-55P_2O_5$ (after aqueous attack).

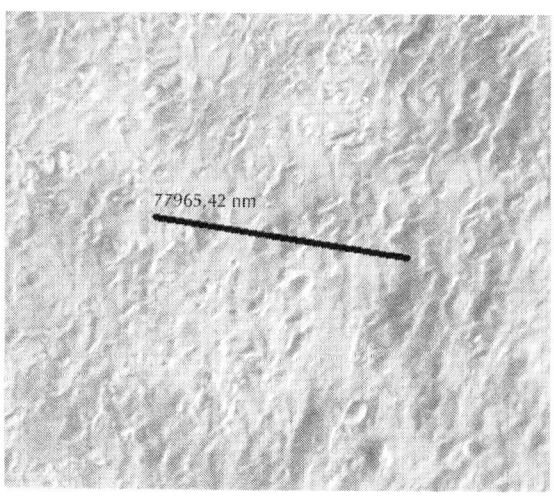

(a)

64 Drilling Engineering

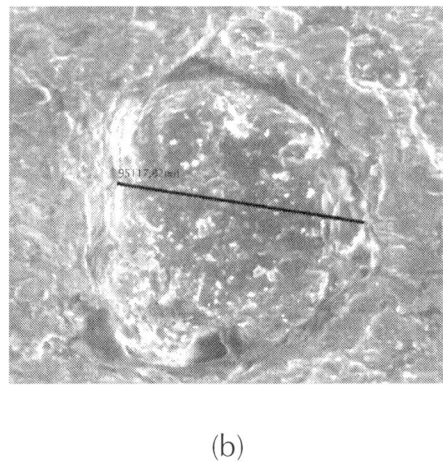

(b)

Figure 4: (a) Microscopic photograph of the surface of the glass $10Na_2O$-$30Fe_2O_3$-$5Cr_2O_3$-$55P_2O_5$, before aqueous attack (Resolution of 50 μm); (b) Microscopic photograph of the surface of the glass $10Na_2O$-$30Fe_2O_3$-$5Cr_2O_3$-$55P_2O_5$, after 22 days of aqueous attack.

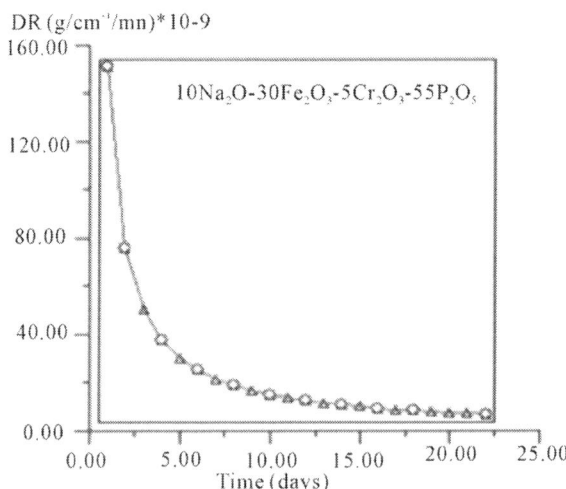

Figure 5: Dissolution rate for sodium chromium iron phosphate glassas a function of immersion time in aqueous solution at 90°C.

The studied glass possesses a strong chemical durability. This can be used in many domains, especially in the vitrification of nuclear wastes which represents a clean alternative to the traditional way of burying hazardous wastes in isolated lands.

ACKNOWLEDGEMENTS

The authors wish to thank National Center for Scientific and Technical Research [Division of Technical Support Unit for Scientific Research (TSUSR) Rabat, Morocco] for helpful discussions. The present work is supported by the Laboratory of Physics of High Energy and Condensed Matter, and Laboratory of Inorganic Chemistry (University Hassan II, Faculty of Sciences Ain Chock).

REFERENCES

1. L. M. Sanford and P. A. Tick, US Patent 4.314.031, 1982.
2. P. Y. Shih, S. W. Yung and T. S. Chin, "Thermal and Corrosion Behavior of P_2O_5-Na_2O-CuO Glasses," Journal of Non-Crystalline Solids, Vol. 224, No. 2, 1998, pp. 143- 152. doi:10.1016/S0022-3093(97)00460-2
3. P. Y. Shih, S. W. Yung, C. Y. Chen, H. S. Liu and T. S. Chin, "The Effect of SnO and $PbCl_2$ on Properties of Stannous Chlorophosphate Glasses," Materials Chemistry and Physics, Vol. 50, No. 1, 1997, pp. 63-69. doi:10.1016/S0254-0584(97)80185-X
4. T. Y. Wei, Y. Hu and L. G. Hwa, "Structure and Elastic Properties of Low-Temperature Sealing Phosphate Glasses," Journal of Non-Crystalline Solids, Vol. 288, No. 1-3, 2001, pp. 140-147. doi:10.1016/S0022-3093(01)00612-3
5. H. Niida, M. Takahashi, T. Uchino and T. Yoko, "Preparation and Structure of Organic-Inorganic Hybrid Precursors for New Type Low-Melting Glasses," Journal of Non-Crystalline Solids, Vol. 306, No. 3, 2002, pp. 292- 299. doi:10.1016/S0022-3093(02)01188-2

6. H. S. Liu, P. Y. Shih and T. S. Chin, "Thermal, Chemical and Structural Characteristics of Erbium-Doped Sodium Phosphate Glasses," Physics and Chemistry of Glasses, Vol. 37, 1996, p. 227.
7. M. R. Reidmeyer, M. Rajaram and D. E. Day, "Preparation of Phosphorus Oxynitride Glasses," Journal of NonCrystalline Solids, Vol. 85, No. 1-2, 1986, pp. 186-203. doi:10.1016/0022-3093(86)90090-6
8. H. Yung, P. Y. Shih, H. S. Liu and T. S. Chin, "Nitridation Effect on Properties of Stannous-Lead Phosphate Glasses," Journal of the American Ceramic Society, Vol. 80, No. 9, 1997, pp. 2213-2220. doi:10.1111/j.1151-2916.1997.tb03110.x
9. J. L. Rygel and C. G. Pantano, "Synthesis and Properties of Cerium Aluminosilicophosphate Glasses," Journal of Non-Crystalline Solids, Vol. 355, No. 52-54, 2009, pp. 2622-2629. doi:10.1016/j.jnoncrysol.2009.09.004
10. P. A. Bingham and R. J. Hand, "Sulphate Incorporation and Glass Formation in Phosphate Systems for Nuclear and Toxic Waste Immobilization," Materials Research Bulletin, Vol. 43, No. 7, 2008, pp. 1679-1693. doi:10.1016/j.materresbull.2007.07.024
11. S. Ray, X. Fang, M. Karabulut, G. K. Marasinghe and D. E. Day, "Effect of Melting Temperature and Time on Iron Valence and Crystallization of Iron Phosphate Glasses," Journal of Non-Crystalline Solids, Vol. 249, No. 1, pp. 1- 16. doi:10.1016/S0022-3093(99)00304-X
12. P. Y. Shih and T. S. Chin, "Preparation of Lead-Free Phosphate Glasses with Low Tg and Excellent Chemical Durability," Journal of Materials Science Letters, Vol. 20, No. 19, 2001, pp. 1811-1813. doi:10.1023/A:1012551603964
13. C. M. Shaw and J. E. Shelby, "Effect of Lead Compounds on the Properties of Stannous Fluorophosphate Glasses," Journal of the American Ceramic Society, Vol. 71, No. 5, 1988, pp. C252-C253. doi:10.1111/j.1151-2916.1988.tb05071.x

14. I. W. Donald, "Preparation, Properties and Chemistry of Glassand Glass-Ceramic-To-Metal Seals and Coatings," Journal of Materials Science, Vol. 28, No. 11, 1993, pp. 2841-2886. doi:10.1007/BF00354689
15. A. Šanti and A. Moguš-Milankovi , "Charge Carrier Dynamics in Materials with Disordered Structures: A Case Study of Iron Phosphate Glasses," Croatica Chemica Acta, Vol. 85, No. 3, 2012. doi:10.5562/cca1989
16. S. Aqdim, H. El Sayouty, B. Elouad and J. M. Greneche, "Chemical Durability and Structural Approach of the Glass Series (40-y) Na_2O-yFe_2O_3-$5Al_2O_3$-$55P_2O_5$-by IR, X-Ray Diffraction and Mössbauer Spectroscopy," IOP Conference Series: Materials Science and Engineering, Vol. 28, 2012, Article ID:012003.
17. B. C. Sales and L. A Batner, "Lead-Iron Phosphate Glass: A Stable Storage Medium for High-Level Nuclear Waste," Sciences, Vol. 226, No. 4670, 1984, pp. 45-48.doi:10.1126/science.226.4670.45
18. B. Kumar and S. Lin, "Redox State of Iron and Its Related Effects in the CaO-P_2O_5-Fe_2O_3 Glasses," Journal of the American Ceramic Society, Vol. 74, No. 1, 1991, pp. 226-228.doi:10.1111/j.1151-2916.1991.tb07322.x
19. S. T. Reis, M. Karabulut and D. E. Day, "Chemical Durability and Structure of Zinc-Iron Phosphate Glasses," Journal of Non-Crystalline Solids, Vol. 292, No. 1-3, 2001, pp. 150-157. doi:10.1016/S0022-3093(01)00880-8
20. T. Jermoumi, M. Hafid and N. Toreis "Density, Thermal and FTIR Analysis of (50-x)BaO.xFe_2O_2·$50P_2O_5$ Glasses," Physics and Chemistry of Glasses—European Journal of Glass Science and Technology Part B, Vol. 43, No. 3, pp. 129-132.
21. S. Aqdim, "Identification et Etudes Thermique et Electrique des Phases vitreuses des Systèmes Ternaires $Li_2OM_2O_3$-P_2O_5 (M= Cr, Fe)," 1990.
22. S. T. Reis, D. L. Faria, J. R. Martinelli, W. M. Pontuschka, D. E. Day and G. S. M. Partiti, "Structural Features of Lead

Iron Phosphate Glasses," Journal of Non-Crystalline Solids, Vol. 304, No. 1-3, 2002, pp. 188-194. doi:10.1016/S0022-3093(02)01021-9

23. S. Cai, W. J. Zhang, G. H. Xu, J. Y. Li, D. M. Wang and W. Jiang, "Microstructural Characteristics and Crystallization of CaO-P_2O_5-Na_2O-ZnO Glass Ceramics Prepared by Sol-Gel Method," Journal of Non-Crystalline Solids, Vol. 355, No. 4-5, 2009, pp. 273-279.doi:10.1016/j.jnoncrysol.2008.11.008

24. R. K. Brow, D. R. Tallant, S. T. Myers and C. C. Phifer, "The Short-Range Structure of Zinc Polyphosphate Glass," Journal of Non-Crystalline Solids, Vol. 191, No. 1-2, 1995, pp. 45-55. doi:10.1016/0022-3093(95)00289-8

25. S. T. Reis, M. Karabulut and D. E. Day, "Structural Features and Properties of Lead-Iron-Phosphate Nuclear Wasteforms," Journal of Nuclear Materials, Vol. 304, No. 2-3, 2002, pp. 87-95. doi:10.1016/S0022-3115(02)00904-2

26. C. R. Rambo, L. Ghussn, F. F. Senc and J. R. Martinelle, "Manufacturing of Porous Niobium Phosphate Glasses," Journal of Non-Crystalline Solids, Vol. 352, No. 32-35, 2006, pp. 3739-3743. doi:10.1016/j.jnoncrysol.2006.03.104

27. H. Doweidar, Y. M, Moustafa, K. EL-Egili and I. Abbas, "Infrared Spectra of Fe_2O_3-PbO-P_2O_5 Glasses," Vibrational Spectroscopy, Vol. 37, No. 1. 2005, pp. 91-96.doi:10.1016/j.vibspec.2004.07.002

28. C. S. Ray, X. Fang, M. Karabulut, G. K. Marasinghe and D. E, Day, "Effect of Melting Temperature and Time on Iron Valence and Crystallization of Iron Phosphate Glasses," Journal of Non-Crystalline Solids, Vol. 249, No. 1, 1999, pp. 1-16. doi:10.1016/S0022-3093(99)00304-X

29. A. Mogus-Milankovic, A. Santic, S. T. Reis, K. Furic and D. E. Day, "Mixed Ion-Polaron Transport in Na_2O-$PbOFe_2O_3$-P_2O_5 Glasses," Journal of Non-Crystalline Solids, Vol. 342, No. 1-3, 2004, pp. 97-109. doi:10.1016/j.jnoncrysol.2004.07.012

30. J.S. Brooks, G.L. Williams and D.W. Allen; Phys. Chem. Glasses 33 (1992) 171.

31. X. Yu, D. E. Day, G. J. Long and R. K. Brow, "Properties and Structure of Sodium-Iron Phosphate Glasses," Journal of Non-Crystalline Solids, Vol. 215, No. 1, 1997, pp. 21-31. doi:10.1016/S0022-3093(97)00022-7
32. S. T. Reis, M. Karabulut and D. E. Day, "Chemical Durability and Structure of Zinc-Iron Phosphate Glasses," Journal of Non-Crystalline Solids, Vol. 292, No. 1-3, 2001, pp. 150-157. doi:10.1016/S0022-3093(01)00880-8
33. G. Malow, W. Lutze and R. C. Ewing, "Alteration Effects and Leach Rates of Basaltic Glasses: Implications for the Long-Term Stability of Nuclear Waste Form Borosilicate Glasses," Journal of Non-Crystalline Solids, Vol. 67, No. 1-3, 1984, pp. 193-205. doi:10.1016/0022-3093(84)90156-X

Chapter 4

A Comparison on Core Drilling of Silicon Carbide and Alumina Engineering Ceramics with Mono-layer Brazed Diamond Tool using Surfactant as Coolant

F.L. Zhang, P. Liu, L.P. Nie, Y.M. Zhou, H.P. Huang, S.H. Wu, and H.T. Lin

School of Mechanical and Electronic Engineering, Guangdong University of Technology, Guangzhou 510006, P.R. China

ABSTRACT

Core drilling with diamond tool is an efficient method for machining of holes in engineering ceramics. A comparative study on core drilling of silicon carbide and alumina engineering ceramics with mono-layer brazed diamond tool was carried out. Two surfactant (anionic type 1631 and nonionic type OP10) water solutions were used as coolant in core drilling. The effect of the coolant's type and concentration on drilling torque and drilling efficiency were also investigated. The morphologies of machined surface of ceramics and the wear of diamond grits were examined. According to the results, the anionic surfactant 1631 water solution produces a lower drilling torque and higher drilling efficiency than that of nonionic OP10. Both silicon carbide and alumina demonstrate a ductile regime of material removal under the lower maximum undeformed chip thickness (0.08 mm). The wear of brazed diamond grits for drilling of silicon carbide is much severer than that of alumina.

INTRODUCTION

The superior properties such as higher hardness, thermal and chemical resistance, lower thermal expansion and electrical conductivity make engineering ceramics widely applied in different fields. However, machining of engineering ceramics is a tough work because it would take great cost to attain the tight tolerances and dimensions with the acceptable surface and sub-surface damage. The research on material removal mechanisms and the efforts to reduce the occurrence of the cracks and damages in the machined surface and subsurface of engineering ceramics have been paid much attention in the previous studies [1], [2] and [3].

Drilling is one of the important machining works for some ceramic parts with complex shapes and structures. The core drilling with diamond tool [4] and various non-traditional methods such as laser beam machining (LBM) [4] and [5], rotary ultrasonic machining (RUM) [6], electrical discharge machining (EDM)[7] and [8], electrochemical discharge machining (ECDM) [9], abrasive water jet machining (AWJM) [10]and ion beam machining [11] can be adopted to drill different kinds of holes in ceramics. The weakness of non-traditional methods is the lower material removal rate, the higher energy consumption and cost. Core drilling with diamond tools is still considered to be an ideal option for the machining of the relatively larger holes in engineering ceramics because of the higher efficiency and the cheaper devising [4] and [13].

The studies on core drilling of alumina have been carried out in some literatures. For example, Zhang studied the effect of core drilling parameters on the materials removal rate of alumina by diamond tool [4]. Gao optimized the tool structure, and investigated the axial force and the rotating speed in core drilling of alumina using impregnated diamond tool [13]. Silicon carbide is another popular engineering ceramic with higher hardness (microhardness 23.0 GPa) than alumina (microhardness 18.9 GPa) [14]. However, core drilling of silicon carbide with diamond tool has rarely been studied, except for some non-traditional methods such as laser drilling and ultrasonic drilling [15] and [16].

Owing to the higher protrusion of diamond grits, mono-layer brazed diamond tools own better machining performance compared to sintered and electroplated diamond tools [17], [18] and [19]. Higher bonding strength of diamond grits can also be achieved because of the chemical bonding between the reactive element such as Ti, Cr in the filler alloy and the diamond [20], [21], [22], [23] and [24]. Mono-layer brazed diamond tools have been adopted in high speed grinding of some engineering ceramics [25]. So far comparative study on core drilling of different engineering ceramics with mono-layer brazed diamond tool has rarely been reported. On the other hand, in the previous studies, water is usually used as coolant in core drilling of ceramic [4] and [13] with diamond tool. The influence of surfactant as coolant on the core drilling of engineering ceramics has also not been found. In consideration of environmental friendship, besides of water, simple and green coolants may help to improve the drilling efficiency. So in the present work two types of surfactants, namely anionic type 1631 and nonionic type OP10, have been adopted as coolants in the core drilling. The dependence of specific energy of core drilling of silicon carbide and alumina on maximum undeformed chip thickness has been examined. The effect of the surfactants on the drilling torques and the drilling efficiency will be investigated. The morphologies of the machined surface of ceramics and the wear feature of the brazed diamond grits on the core drill have been observed and analyzed.

EXPERIMENTAL

Mono-layer brazed diamond core drill tools were fabricated by high-temperature brazing in a vacuum furnace (950 °C, 6.67×10^{-3} Pa) with a commercial BNi7 type filler alloy, the tool's photo and dimensions are shown in Figure 1. The diamond grits for brazing is the commercial SDB1125 from Element Six with three mesh sizes of 30/40, 40/50 and 50/60. The silicon carbide and alumina ceramic slabs are obtained from Feng Yuan Weifang Ceramic Co. Ltd., and their dimensions and properties of ceramics are shown in Table 1. The water solutions of commercial surfactants, namely hexadecyl trimethyl ammonium Bromide (anionic type, named 1631) and the alkylphenol polyoxyethylene (nonionic type, named OP10), are prepared to be the coolants of core drilling. The tangential drilling force (Ft) was measured on a high-speed machining center (DMG-60 T) with the spindle speed from 3000–5000 rpm and the feed rate from 0.4–1.6 mm/min using a dynamometer (YDM-I97, Piezoelectric) as illustrated in Figure 2. The drilling efficiency is carried out on a drilling machine (YG1-1D) under the axial load of 20 Kg with the spindle speed of 3850 rpm, and the drilling time for one hole was recorded by a stopwatch. The wear of diamond grits and the surface morphologies of the drilled ceramics were observed by a scanning electron microscopy (SEM, JEOL JED- 2300).

Figure 1: Photo of mono-layer brazed diamond core drill tool.

Table 1: Dimensions and properties of silicon carbide and alumina ceramics slabs

	Dimension	Density	Purity	Hardness	Toughness
	(mm)	(g/cm^3)	(%)	(GPa)	(MPa.m$^{1/2}$)
Silicon carbide	50×50×8	3.1	>97	23	4.0
Alumina	50×50×8	3.88	>99	18.9	3.5

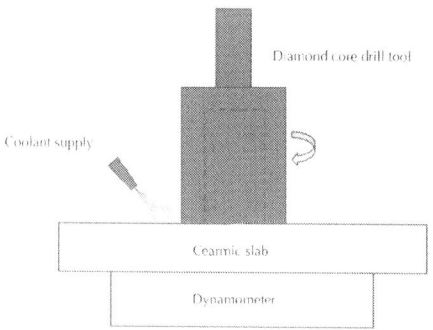

Figure 2: Schematic illustration of the measuring of tangential drilling force.

RESULTS AND DISCUSSION

Specific Energy for Drilling Silicon Carbide and Alumina

The specific energy is a fundamental parameter to characterize the grinding related process, as well as other machining processes, which can manifest the consequence of the prevailing mechanisms of abrasive/workpiece interactions. In the core drilling of ceramics with diamond tool, the specific energy u can be obtained from the relationship [26]:

$$u = F_t v_s / Q_w \tag{1}$$

Where F_t is the tangential force of drilling which can be measured by the dynamometer, v_s is the velocity of core drilling, Q_w is the volumetric removal rate. The maximum undeformed chip thickness h_m can be considered as the feed in axial direction in a revolution cycle as schematically illustrated in Figure 3, which is written as follow:

$$h_m = v_f / \omega \tag{2}$$

Where v_f is the feed rate, ω is the spindle speed. The dependences of specific energy u on the maximum undeformed chip thickness h_m for silicon carbide and alumina are shown in Figure 4.

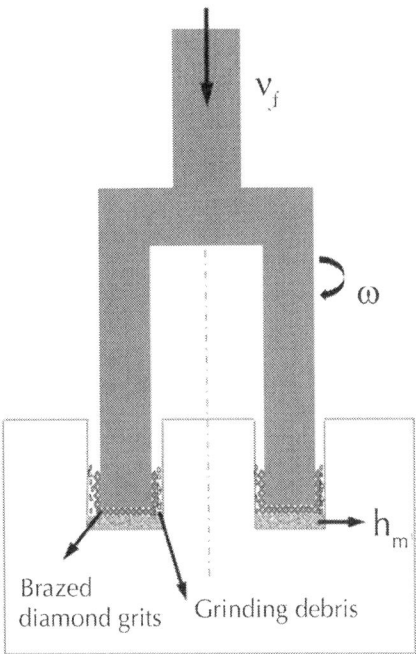

Figure 3: Schematic illustration of the maximum undeformed chip thickness in core drilling.

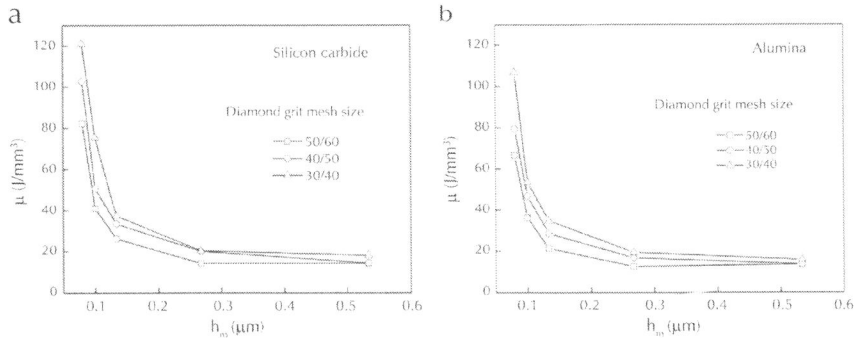

Figure 4: The dependence of specific energy u on the maximum undeformed chip chickness hm in core drilling of silicon carbide (a) and alumina (b).

It is found that with the decreasing of the maximum undeformed chip thickness h_m the specific energy u is increased for both silicon carbide and alumina. This inverse relationship between u and h_m can be referred to the size effect [27]. The increased u at very low h_m has been attributed to an increased tendency for ductile plowing rather than fracture as the abrasive grains interact with the hard and brittle material.

The critical depth of cut, d_c, for a brittle-ductile transition in machining brittle materials given by Bifano is shown below [28]:

$$d_c = \beta(E/H)(T/H)^2 \tag{3}$$

Where β is related to the geometry of the tool and is usually taken to be 0.15, E is elastic modulus, H hardness and T toughness. The value of dc of silicon carbide and alumina is found to be 0.06 and 0.11 μm, respectively. In this study the lowest value of h_m is 0.08 μm, which is quite close to the d_c of silicon carbide and alumina. Thus the plowing with the ductile mode of material removal can be considered to account for the increased specific energy of core drilling in Figure 4. On the other hand, the finer diamond grits also lead to a higher specific energy, which may related to the semi-included angle of diamond grits. The diamond grits with mesh size of 50/60 own lower semi-included angle than those with mesh size of 40/50 and 30/40. The specific energy of ductile plowing u_p can be expressed as follow [27]:

$$u_p/H = a_p/(90 - q) + b_p \tag{4}$$

where H is the hardness of brittle material, a_p and b_p are the fitting constants, respectively. The coarser diamond grits with higher can result in larger specific energy of plowing. The specific energy for core drilling of silicon carbide is slightly higher than that that of alumina at the lower maximum undeformed chip thickness h_m. This can also be explained by Eq. (4). Owing to the higher hardness H, a higher specific energy of ductile plowing u_p is needed for drilling silicon carbide than that for alumina.

Effect of Surfactant on the Core Drilling Torques and Drilling Efficiency

The drilling torque can be obtained by multiplying the tangential force F_t by the radius of the core drill bit, which reflects the resistance of core drilling. For the convenience of comparison the machining center's spindle speed and the feed rate are controlled as 5000 rpm and 0.4 mm/min (h_m 0.08 mm), respectively. In Figure 5 we can found that the concentration of the surfactant water solutions can remarkably influence the drilling torques of silicon carbide and alumina. The drilling torque drops with the increase of the coolant's concentration within the range of 0(water)-5%. In the previous studies on core drilling of rocks, Mills and Westwood consider that the chemomechanical effect (CME) of surfactant contributes to the increase of drilling efficiency and the wear reduction of diamond grits [29]. Macmillan found that the surfactant can markedly affect the friction behavior of MgO and soda lime glass by influencing the near surface flow properties (i.e. microhardness) [30]. The surfactant solution can alter the microhardness of the non-metals by changing its zeta potential (-potential). The zeta potential is the potential difference across phase boundaries between solids and liquids. It is an index of the magnitude of the electrostatic repulsive interaction between particles. The characteristics of the solid-liquid interface are also influenced by zeta potential [31]. When zeta potential is close to zero, the mobility of dislocation and the coefficient of friction can be reduced at the same time [30]. Thus the reduction of core drilling torques in Figure 5 can be explained by the chemomechanical effect (CME) of surfactant. This means that the anionic 1631 and the nonionic OP10 water solution could have induced the reduction of the microhardness of ceramics and the coefficient of friction between diamond grits and ceramics When comparing anionic type 1631 with the nonionic OP10, it is found that the drilling torques with the 1631 are lower than that with OP10 as shown in Figure 5. The capability of changing zeta potential of silicon carbide and alumina with 1631 could be higher than that with OP10. Nevertheless, the detail mechanism of the effect of

1631 and OP10 on the zeta potential and the microhardness of silicon carbide and alumina need further investigation.

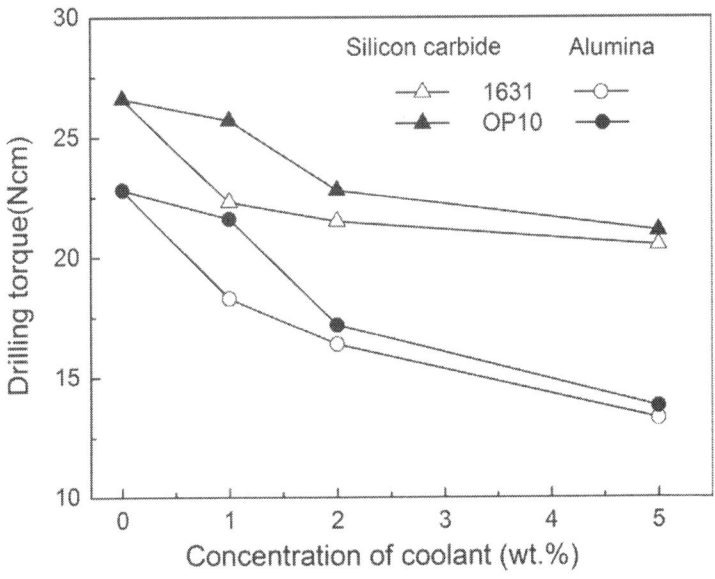

Figure 5: The dependence of drilling torques on the concentration of OP10 (a) and 1631(b).

As illustrated in Figure 6 and Figure 7, it is clearly shown that silicon carbide owns much lower drilling efficiency than alumina. The higher hardness of silicon carbide produces the higher drilling torque as discussed above, and accordingly the drilling efficiency is reduced. On the other hand, the drilling efficiency is increased slightly with the increase of the concentration of surfactants, which is also caused by the chemomechanical effect (CME) of surfactant as discussed above. On the other side, the core drill tool with coarser diamond grits (30/40 mesh) owns higher drilling efficiency than that with finer grits (40/50 and 50/60 mesh). Higher protrusion of diamond grits can lead to higher sharpness of diamond tool which produces higher drilling efficiency.

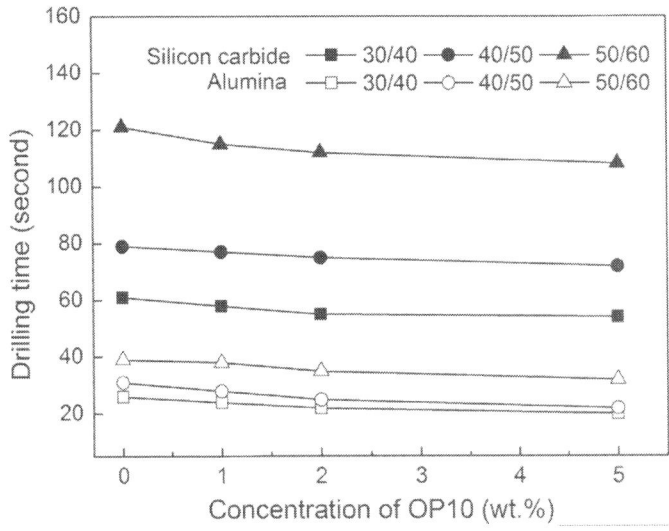

Figure 6: The dependence of drilling time of one hole in silicon carbide (a) and alumina (b) on the concentration of OP10.

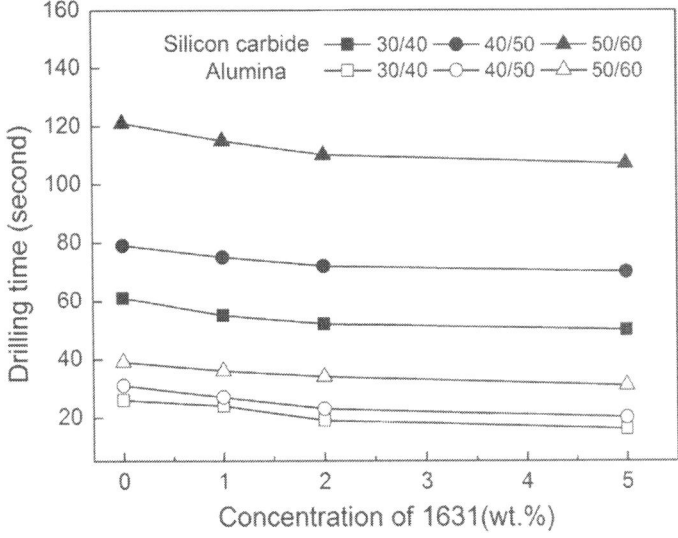

Figure 7: The dependence of drilling time of one hole in silicon carbide(a) and alumina(b) on the concentration of 1631.

Surface Morphology of Drilled Silicon Carbide and Alumina Ceramics

For observing the morphology of drilled surface of ceramics, the drilled ring grooves in the ceramic slabs were obtained with the drilling depth of 1.6 mm, and the surface characteristics of the ring grooves of silicon carbide and alumina are shown in Figure 8. The inner edge (labeled by letter A) of the ring grooves own more chipping than the outer edge (labeled by letter C) as pointed by the dark arrows. This chipping at the inner edge could be the result of different cooling effect between inner and outer part of ring grooves in the core drilling. Normally coolant tends to flow to the outer part of ring due to the driving of the centrifugal force, thus the poor cooling effect at the inner edge may cause higher thermal stress and produce more cracks and chipping.

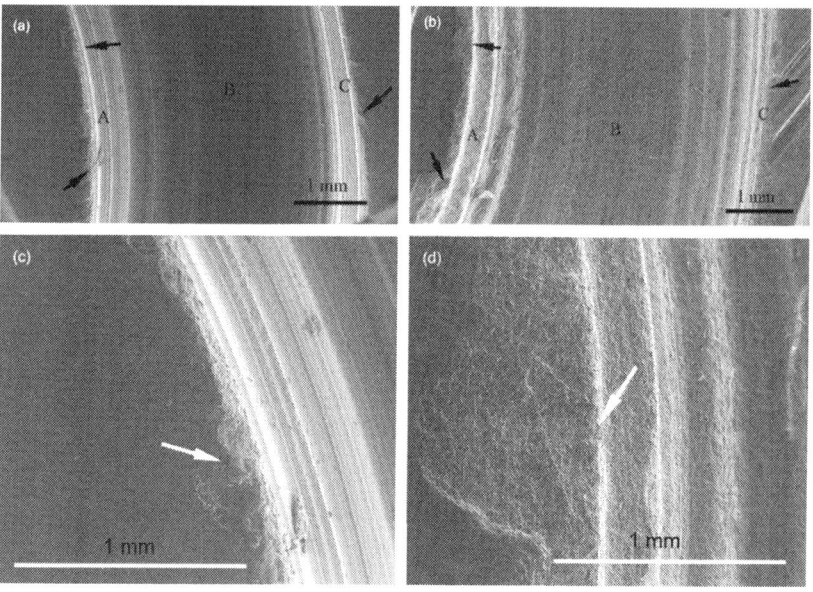

Figure 8: SEM images of the drilled ring groove and the edge chipping of inner ring groove of silicon carbide (a,c) and alumina (b,d).

It can also be found that the chipping area in the inner edge of alumina is much larger than that of silicon carbide as pointed by the bright arrows in Figure 8 (c) and (d). The fracture toughness of silicon carbide and alumina is similar (3.5 and 4.0 MPa.m$^{1/2}$) [30]. We assume that the larger chipping area at the ring grooves of alumina may be caused by the thermal stress of ceramics in the core drilling process. Silicon carbide owns much higher thermal conductivity (115–150 W/mK) than alumina (16–31 W/mK) [32]. Even if the supply of coolant is enough, the drilling area of ceramic may not be cooled sufficiently due to the effect of centrifugal force. Lower thermal conductivity leads to higher thermal stress in the drilling area, which induce the larger chipping area in alumina than that in silicon carbide.

Figure 9 presents the surface morphology of the bed area of ring groove which labeled by letter B in Figure 8. As shown in Figure 9, a dominated ductile regime of material removal can be found in both silicon carbide and alumina, which is in correspondence with the brittle-ductile transition criterion proposed by Bifano [28] and discussed in 3.1. However, recently Li has found that the brittle-ductile transition criterion can be affected by the machining speed, and in high speed grinding the penetration depth of grit can be greatly enlarged without fracture cracks generation [33]. In the present study, the machining velocity of core drilling is in the range of 3.14–5.23 m/s, which is much lower than the velocity 45–60 m/s which is the transition of high speed grinding. At the low machining speed, the ductile material removal in this core drilling is mainly induced by the aforementioned lower maximum undeformed chip thickness h_m.

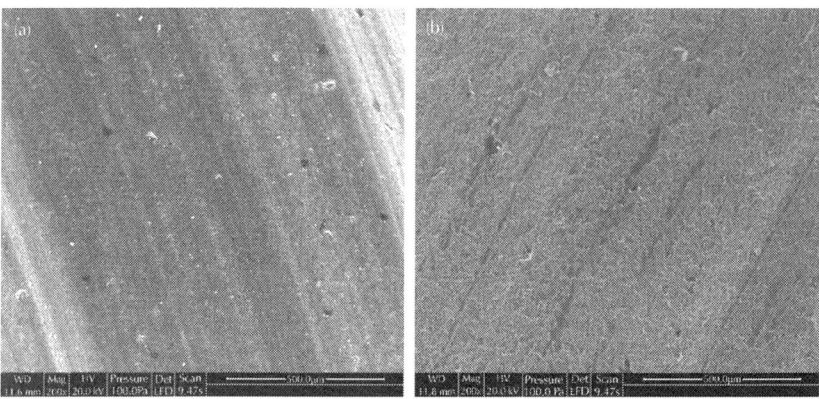

Figure 9: Surface morphology of the bed area of drilled ring groove of silicon carbide (a) and alumina (b).

The Wear of Brazed Diamond Core Drilling Tool

In order to examine the wear of brazed diamond core drilling tools, one hole was drilled by the brazed diamond core drill tools on silicon carbide and alumina ceramics, respectively. And the core drilling is carried out on the high-speed machining center with the controlled parameters and conditions (spindle speed of 5000 rpm, feed rate of 0.4 mm/min, coolant of 2% 1631 water solution, 40/50 mesh size of diamond grits). The as-brazed diamond tools and the used diamond tools were examined by SEM as presented in Figure 10. It can be seen that the diamond grits on the as-brazed tool are well hold and wetted by the brazed alloy with high protrusions. After the core drilling of one hole on silicon carbide, the wear of diamond grits is much severer than that on alumina as illustrated in Figure 10(c,d) and (e,f). The macro-fracture dominates the wear of diamond grits in drilling of silicon carbide as seen in Figure 10(c) and (d). Some flattened and polished diamond grits (labeled by letter B) and some pull-out craters of diamond grits (marked by letter C) can also be located. However, according to the diamond grits for drilling of alumina, very few macro-fracture marked by bright arrows and some micro-fracture marked by dark arrows can be seen in Figure 10(e,f). The severer wear of diamond grits in the core drilling of silicon carbide can also be attributed to its higher hardness, which also induces the higher drilling torque as discussed in 3.2.

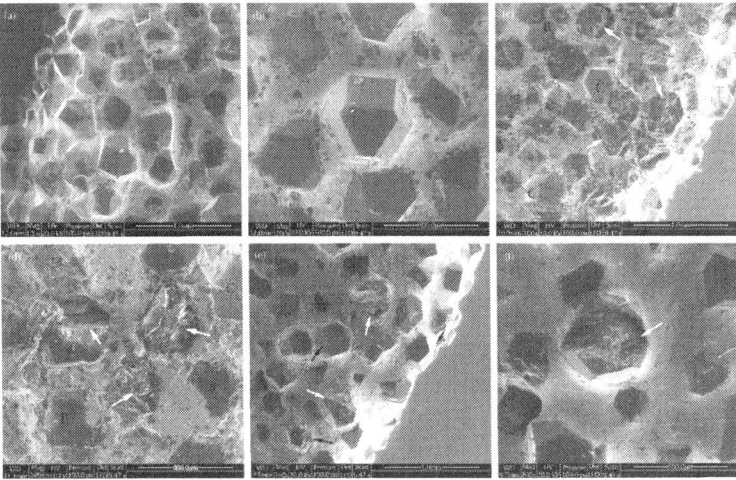

Figure 10: Surface morphology of as-brazed diamond tool (a, b) and the wear characteristics of tool after drilling one hole in silicon carbide (c, d) and alumina (e,f).

CONCLUSIONS

In the core drilling of silicon carbide and alumina engineering ceramics by mono-layer brazed diamond tool, a size effect that the specific energy of core drilling is increased with the decrease of undeformed chip thickness can be found. The drilling torques can be remarkably reduced with the increase of the concentration of surfactants OP10 and 1631 because of the chemomechanical effect (CME). The anionic type surfactant 1631 water solution can produce lower drilling torque than the nonionic type OP10. The brazed diamond tools confront higher drilling torques in drilling of silicon carbide than that of alumina ceramic due to the higher hardness of silicon carbide. The core drilling efficiency of the monolayer brazed diamond tool for silicon carbide is lower than that for alumina. Under the lower undeformed chip thickness (0.08 mm), a ductile regime of material removal dominates the core drilling of silicon carbide and alumina. At the drilled inner ring grooves of alumina the edge chipping area is larger than that of silicon carbide. The macro-fracture and dislodgement of diamond grits characterize the wear of the brazed diamond tool in core drilling of silicon carbide, which is much severer than that for alumina.

ACKNOWLEDGMENTS

The authors would like to acknowledge the financial support from the National Natural Science Foundation of China (NSFC) with Grant No. 51275096.

REFERENCES

1. K. Suzuki, T. Uematsu, S. Mishiro, A new grinding method for ceramics using a biaxially vibrated non-rotational ultrasonic tool Ann. CIRP, 42 (1993), pp. 375–378
2. K.P. Rajurkar, Z.Y. Wang, A. Kuppattan Micro removal of ceramicmaterial (Al2O3) in the precision ultrasonic machining Precis. Eng., 23 (1999), pp. 73–78
3. K. Ueda, T. Sugita, H. Hiraga A J-integral approach to material removal mechanisms in microcutting of ceramics Ann. CIRP, 40 (1991), pp. 61–64
4. Q.H. Zhang, J.H. Zhanga, D.M. Sun, G.D. Wang Study on the diamond tool drilling of engineering ceramics J. Mater. Process.Technol., 122 (2002), pp. 232–236
5. N. Nedialkov, M. Sawczak, R. Jendrzejewski, P. Atanasov, M. Martin, G.S. liwinski Analysis of surface and material

modifications caused by laser drilling of AlN ceramics Appl. Surf. Sci., 254 (2007), pp. 893–897
6. E. Kacara, M. Mutlua, E. Akmana, A. Demira, L. Candana, T. Canel, V. Gunay, S.ı. nmazcelik Characterization of the drilling alumina ceramic using Nd:YAG pulsed laser J. mater. process. technol, 209 (2009), pp. 2008–2014
7. W.M. Zeng, Z.C. Li, Z.J. Pei, C. Treadwell Experimental observation of tool wear in rotary ultrasonic machining of advanced ceramics, Int. J Mach. Tools Manufact, 45 (2005), pp. 1468–1473
8. Y. Liu, R. Ji, Q. Li, L. Yu, X. Li Electric discharge milling of silicon carbide ceramic with high electrical resistivity, Int. J Mach. Tools Manuf, 48 (2008) (2008), pp. 1504–1508
9. B. Bhattacharyya, B.N. Doloi, S.K. Sorkhel Experimental investigations into electrochemical discharge machining (ECDM) of non-conductive ceramic materials J. Mater. Process. Technol., 95 (1999), pp. 145–154
10. Akkurt The effect of material type and plate thickness on drilling time of abrasive water jet drilling process Mater. Design, 30 (2009), pp. 810–815
11. D.M. Allen, P. Shore, R.W. Evans, C. Fanara, W.O.' Brien, S. Marson, W.O.' Neill, I.o.n. Beam Focused Ion Beam, and Plasma Discharge Machining CIRP Ann. Manuf. Technol, 58 (2009), pp. 647–662
12. S.T. Chen, Z.H. Jiang, Y.Y. Wu, H.Y. Yang Development of a grinding–drilling technique for holing optical grade glass I. J. Mach. Tools Manuf, 51 (2011), pp. 95–103
13. C. Gao, J. Yuan Efficient drilling of holes in Al_2O_3 **armor ceramic using impregnated diamond bits** J. Mater. Process. Tech (2011), pp. 1719–1728
14. Han Huang Ling Yin Ceramic Response to High Speed Grinding Mach. Sci.Technol., 8 (2004), pp. 21–37
15. A.N. Samant, N.B. Dahotre Laser machining of structural ceramics-A review J. Eur. Ceram. Soc., 29 (2009), pp. 969–993

16. H. Hocheng, N.H. Tai, C.S. Liu Assessment of ultrasonic drilling of C/SiC composite material Compos. Part A Appl. Sci. Manuf, 31 (2000), pp. 133–142
17. H.E. Hintermann, A.K. Chattopadhyay New generation superabrasive tool with monolayer configuration Diamond and Related Materials, 1 (1992), pp. 1131–1143
18. C.M. Sung Brazed diamond grid: a revolutionary design for diamond saws Diamond Relat. Mater, 8 (1999), pp. 1540–1543
19. Trenker, H. Seidemann High-vacuum brazing of diamond tools Ind.diamond rev, 62 (2002), pp. 49–51
20. A.K. Chattopadhyay, L. Chollet, H.E. Hintermann\ Experimental investigation on induction brazing of diamond with Ni-Cr hard facing alloy under argon atmosphere J. Mater. Sci., 26 (1991), pp. 5093–5100
21. A.K. Chattopadhyay, L. Chollet, H.E. Hintermann Induction brazing of diamond with Ni-Cr hardfacing alloy under argon atmosphere Surf.Coat. Technol, 45 (1991), pp. 293–298
22. F.A. Khalid, U.E. Klotz, H.-R. Elsener, B. Zigerlig, P. Gasser On the interfacial nanostructure of brazed diamond grits Scripta Mater., 50 (2004), pp. 1139–1143
23. S.F. Huang, H.-L. Tsai, S.-T. Lin Effects of brazing route and brazing alloy on the interfacial structure between diamond and bonding matrix Mater.Chem. Phys., 84 (2004), pp. 251–258
24. U.E. Klotz, F.A. Khalid, H.R. Elsener Nanocrystalline phases and epitaxial interface reactions during brazing of diamond grits with silver based Incusil-ABA alloy Diamond Relat. Mater, 15 (2006), pp. 1520–1524
25. J.Y. Chen, X.P. Xu The temperatures and energy partitions for high speed grinding of alumina with a brazed diamond wheel Mach.Sci. and Technol, 14 (2010), pp. 440–454
26. S. Malkin Grinding Technology: Theory and Application of Machining with Abrasives Wiley, New York (1989)

27. T.W. Hwang, C.J. Evans, S. Malkin Size effect for specific energy in grinding of silicon nitride Wear, 225–229 (1999), pp. 862–867
28. T.G. Bifano, T.A. Dow, R.O. Scattergood Ductile-regime: a new technology for machining brittle materials ASME J. Eng. Industry, 113 (1991), pp. 184–189
29. J.J. Mills, A.R.C. Westwood Influence of chemomechanically active fluids on diamond wear during hard rock drilling J. Mater. Sci., 13 (1978), pp. 2712–2716
30. N.H. Macmillan, R.D. Huntington, A.R.C. Westwood Chemomechanical control of sliding friction behaviour in non-metals J. Mater. Sci., 9 (1974), pp. 697–706
31. B.B. Weiner, W.W. Tscharnuter, D. Fairhurst.Zeta potential: a new approach Proceedings of the Canadian Mineral Analysts Meeting (1993)
32. The Ceramic Society of Japan, Advanced Ceramic Technologies & Products, Springer Verlag, 2012
33. B.Z. Li, J.M. Ni, J.G. Yang, S.Y. Liang Study on high-speed grinding mechanisms for quality and process efficiency Int J Adv Manuf Technol, 70 (2014), pp. 813–819

Chapter 5

Nature of Drilling Forces During Spark Assisted Chemical Engraving

Jana D. Abou Ziki and Rolf Wüthrich

Department of Mechanical & Industrial Engineering, Concordia University, 1455 de Maisonneuve Blvd. West, Montreal H3G 1M8, Canada

ABSTRACT

Tool–substrate contact drilling forces proved to be promising for monitoring the SACE drilling progress. Although these forces were recently measured, their nature was never studied. This is investigated in the present work where it is found that a tool–glass bond is created while drilling. Preliminary investigations demonstrated that this bond does not break due to specific temperature drop and that it is formed just before tool-glass contact occurs. The bond is therefore expected to be chemically formed where more investigation is needed to further clarify this. The present results add up to the knowledge regarding the SACE machining mechanism.

INTRODUCTION

Spark assisted chemical engraving (SACE) machining is suited for non-conductive materials, mainly glass, where its main mechanism is believed to be high temperature etching [1]. The substrate is machined in an electrochemical cell where both the tool and the counter-electrodes are dipped in an electrolytic solution. Machining occurs due to generation of high energy discharges from the tool, upon application of high enough voltage, to the substrate through a gas film formed around the tool tip. In fact, several studies [2],[3], [4] and [5] revealed that the local temperature can reach 500–600°C.

During this process, machining occurs when the tool is in close vicinity to the substrate (less than 25 μm separation distance [6] and [7]). For constant velocity-feed drilling, the tool is moved towards the glass work-piece at a constant feed-rate, while drilling, where tool–glass contact may result [7]. Drilling is also done in gravity-feed configuration where the tool is pushed into the substrate by a constant force. In this case, both the tool and the glass are always in contact during drilling [8].

Recently, studies were done on measuring the tool–substrate contact force during SACE drilling in a step towards using it for monitoring the drilling progress. This was done by means of force sensors either located in the machining head where the tool is mounted [9] or fixed beneath the substrate [10]. These forces have also been characterized for several machining conditions in [9]. However, their nature is not yet known.

In this letter, the nature of SACE drilling forces is investigated. It is demonstrated that a tool–glass bond can be formed while drilling. The present findings show that this bond is most likely formed chemically, similar to anodic bonding, where further investigation is needed to clarify this.

EXPERIMENTAL METHODS

The experimental set-up is composed of a machining head mounted on a Z-stage and a processing cell fixed to an XY-stage as described in [9]. The machining head holds the tool–electrode (stainless steel, 500 microns in diameter) and is formed of a force sensor that records forces based on the optically measured deflection of an elastic element of known stiffness ($K = 0.05$ N/μm). The sensor can measure forces causing upward and downward deflection (positive and negative forces respectively). The holes, 100 μm deep, are machined on glass substrate in 30 wt% NaOH solution while moving the tool at a constant feed-rate of 10 μm/s and applying a voltage of 33 V. Whenever the contact force exceeds 0.1 N, the tool motion is stopped until the force disappears after which its motion is restored. Once drilling is done, the tool is kept still for 10 s to allow cooling down the machining zone [11].

DISCUSSION

Results showed that when a contact force exists in the end of drilling (close to 100 μm depth), it sometimes decays to a negative value when the machining voltage is switched off (Fig. 1a). The force occurring just before drilling ends is called Pressing-Force in this text and the decay in the force after switching off the machining voltage is referred to as Delta-Force. The exponential decay in the force is due to local cooling when the voltage is switched off, causing the tool to retract. Upon its retraction, the tool moves upward away from the glass surface. In the example shown in Fig. 1a, the force decays reaching a negative value after which it goes back to zero. This implies that there is a downward deflection such that the tool is pulled towards the glass during a certain time, hence is bonded to it, and is then released causing the force to go back to zero. The force needed to detach the tool from the glass surface is denoted as Glue-Force in this text. Note that when the Pressing-Force is larger than Delta-Force, caused by the maximal tool retraction, the force decays into a positive value as shown in Fig. 1b. In this case, the tool–glass bonding cannot be seen through the force signal.

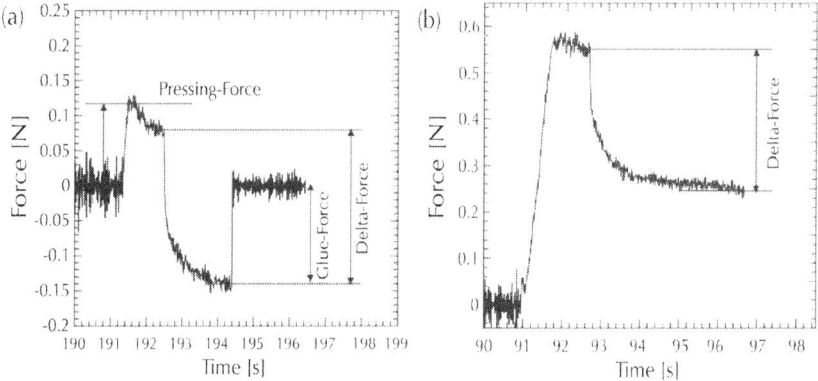

Figure 1: The two forms that the force signal vs. machining time can take upon tool–glass bonding. The Pressing-Force is the force pressing the tool against the glass surface and Delta-Force is the force signal decay after switching off the machining voltage at 192.5 s in case (a) and at 92.8 s in case (b) before the tool is released. The Glue-Force is the force needed to detach the tool from the glass surface.

The effect of the time during which the Pressing-Force is applied, before the voltage is switched off, on the formation of tool–glass bonding is examined. This time is denoted as Pressing-Time as shown in the inset of Fig. 2. The Delta-Force is recorded for various Pressing-Times ranging from 0.1 s (fastest time that could be established due to set-up limitations) up to 1 s. For each Delta-Force, the tool retraction is calculated knowing the force sensor stiffness. This allows estimating the tool temperature based on the thermal model developed in [5]. Results showed that for all Pressing-Times, the Delta-Force is around 0.22 N which corresponds to a tool retraction of 4.5 µm (Fig. 2) implying a tool temperature drop of about 275 °C. As even after 0.1 s Pressing-Time the tool–glass bonding could be observed, this must be a very fast process.

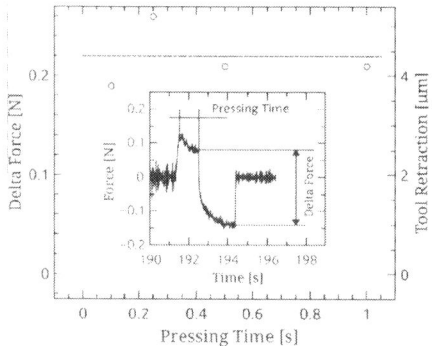

Figure 2: Change in Delta-Force in function of Pressing-Time for 100 μm deep holes machined using 30 wt% NaOH solution, 10 μm/s tool feed-rate and 33 V. The figure inset is a representative example of the recorded force signal in function of time and it depicts the Pressing-Time which is the time during which the Pressing-Force is applied before switching off the voltage.

The amplitude of the Glue-Force for a range of Pressing-Force from 0.05 to 0.25 N is investigated (Fig. 3a). Results show that the Glue-Force is constant, around 0.15 N, for all the Pressing-Force values applied (50 holes drilled for each Pressing-Force). This shows that the tool–glass bond breaks after pulling the tool by a constant force indicating that the bond has a well-defined mechanical strength.

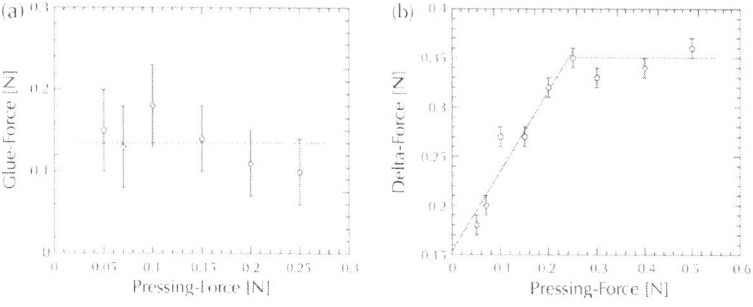

Figure 3: (a) The Glue-Force and (b) Delta-Force in function of the Pressing-Force.

If this bond is formed by solidification of local material beneath the tool, like electrolyte molten salt, upon reduction of local temperature it must break after a specific temperature drop causing the material to dissolve. Fig. 3b shows that Delta-Force, which is proportional to the tool retraction is not constant. This implies that the tool temperature at the moment the bond breaks is variable since for all conditions the initial tool temperature is the same (constant machining voltage is applied). Thus, the bond does not break at a specific temperature and it is not created by solidification of material beneath the tool.

For very large Pressing-Force (higher than 0.25 N in this example), Delta-Force reaches a maximal value (around 0.35 N) corresponding to the maximal possible tool retraction (for the specific applied machining voltage) calculated to be around 8 μm which indicates a temperature of 475 °C. When the data is extrapolated to zero Pressing-Force, a Delta-Force of 0.15 N results. This indicates that the bond forms even when almost no mechanical force is exerted on the tool. It has to be noted that this particular value of Delta-Force corresponds to the Glue-Force found in Fig. 3a. This suggests that the bond is formed chemically, probably similar to field-assisted bonding between metal and glass which occurs at temperatures lower than the glass softening point (less than 700 °C) and during a short time [12]. In the case of SACE, the high electrical field existing in the vicinity of the tool causes a migration of ions from the glass bulk to its surface hence creating a bond between the metallic tool and the glass.

The inspection of the whole bottom surfaces for the various applied Pressing-Force levels, revealed different surface patterns. Typical whole bottom surfaces are depicted in Fig. 4 for each Pressing-Force. It is noticed that for high Pressing-Forces (above 0.25 N) the tool in imprinted on the hole bottom surface which is circular and smooth. For smaller Pressing-Forces, the machined surface is deformed and less circular. This can be explained by the fact that for small Pressing-Force, the tool detaches from the surface before the glass cools down (force jumps from a negative value to zero as seen in Fig. 1a). This causes the glass surface layer to deform. However, for Pressing-Force higher than the fore decay caused by tool retraction, the tool stays in contact with the glass surface (positive force level) during 10 s after the tool retracts as inFig. 1b. In this case the glass is etched around the tool while less etching occurs beneath it where the hot surface layer is imprinted by the tool.

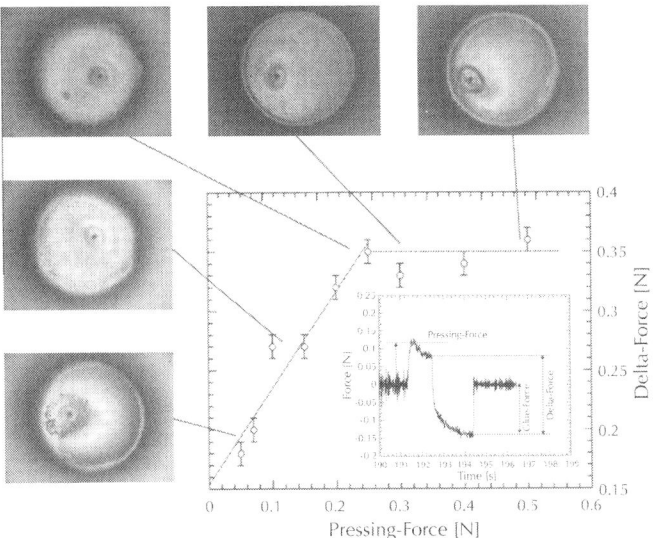

Figure 4: A plot of Delta-Force in function of the Pressing-Force and the corresponding hole bottom surfaces for selected Pressing-Force. The figure inset is a representative example of the recorded force signal in function of time depicting the Pressing-Force, the Glue-Force and the Delta-Force. For higher Pressing-Force, Delta-Force increases until reaching a certain limit dictated by the maximal possible tool retraction for the applied machining conditions.

The present findings explain the limited machining which randomly occurs during gravity-feed and constant velocity-feed drilling. In fact, this may be caused by the tool–glass bonding occurring during certain instants. This explains the stair-case evolution of the hole depth versus drilling time during gravity-feed drilling [8] where machining stops whenever the tool bonds to the glass and proceeds upon tool–glass detachment.

CONCLUSIONS

In this letter, the nature of the tool–substrate contact forces during SACE drilling is studied. Results show that a bond is created between the tool and the glass surface while machining. Preliminary investigations demonstrated that this bond is not caused by the local temperature decay that may cause solidification of the local molten electrolyte or the machined material beneath the tool. In fact, it is found that the bond has a well-defined mechanical strength and can occur at almost zero mechanical tool–glass contact, suggesting that a chemical bond is formed. The bonding process is similar to the one in field assisted bonding.

The present findings enhance the knowledge about SACE machining mechanism. The tool–surface bonding may explain the reduced machining rate occurring for certain durations in gravity-feed or constant-velocity-feed drilling. Moreover, this work provides a method to establish tool–glass bonding in a controlled manner while machining. The current knowledge can be applied to control the surface texture of machined holes, specifically to obtain smoother machined surfaces.

ACKNOWLEDGEMENTS

This work was supported by the Natural Sciences and Engineering Research Council of Canada (NSERC). J.D.A.Z. would like to thank the Ministère de l'Education, du Loisir et du Sport du Quebec (MELS) for the bourse d'excellence pour étudiants étrangers (V1) and Posalux SA for the Posalux Excellence Scholarship.

REFERENCES

1. Wüthrich R, Fascio V. Machining of non-conductive materials using electrochemical discharge phenomenon-an overview. Int J Mach Tools Manuf 2005; 45:1095–108.

2. Kellogg HH. Anode effect in aqueous electrolysis. J Electrochem Soc 1950;97:133–42.
3. Reghuram V. Electrical and spectroscopic investigations in electrochemical discharge machining Ph.D. dissertation. Madras: Indian Institute of Technology; 1994.
4. Jalali M, Maillard P, Wu″thrich R. Toward a better understanding of glass gravity-feed micro-hole drilling with electrochemical discharges. J Micromech Microeng 2009;19:45001–8.
5. Abou Ziki JD, Wu″thrich R. Tool wear and tool thermal expansion during micro-machining by spark assisted chemical engraving. Int J Adv Manuf Technol 2012;61:481–6.
6. Fascio V, Wu″thrich R, Viquerat D, Langen H. 3d microstructuring of glass using electrochemical discharge machining (ECDM). In: Proceedings of 1999 International Symposium on Micromechatronics and Human Science, Nagoya; 1999. p. 179–3.
7. Wu″thrich R. Micromachining using electrochemical discharge phenomenon: fundamentals and applications of spark assisted chemical engraving. New York: William Andrew; 2009.
8. Wu″thrich R, Spaelter U, Wu Y, Bleuler H. A systematic characterization method for gravity-feed micro-hole drilling in glass with spark assisted chemical engraving (SACE). J Micromech Microeng 2006;16(9):1891.
9. Abou Ziki JD, Wu″thrich R. Forces exerted on the tool–electrode during constant-feed glass micro-drilling by spark assisted chemical engraving. Int J Mach Tools Manuf 2013; 73:47–54.
10. Cao XD, Kim BH, Chu CN. Micro-structuring of glass with features less than 100 lm by electrochemical discharge machining. Precis Eng 2009; 33:459–65.
11. Abou Ziki JD. Spark assisted chemical engraving: a novel approach for quantifying the machining zone parameters using drilling forces Ph.D. dissertation.. Canada: Concordia University; 2014.

12. Wallis G, Pomerantz DI. Field assisted glass-metal sealing. J Appl Phys 1969; 40:3946.

Chapter 6

Complexity in Semiconductor Manufacturing, Activity of Antimicrobial Agents, and Drilling of Hydrocarbon Wells: Common Themes and Case Studies

Michael Nikolaou[a], Pratik Misra[a], Vincent H. Tam[b], and Andrew D. Bailey III[c]

[a]Chemical Engineering Department, University of Houston, Houston, TX 77204-4004, USA
[b]University of Houston College of Pharmacy, Houston, TX, USA
[c]Lam Research Corporation, Fremont, CA, USA

ABSTRACT

Complexity manifests itself in various ways and in systems from seemingly disparate areas. In this work we present a sample of complex phenomena that may arise in systems from three different areas: semiconductor manufacturing, activity of antimicrobial agents, and drilling of hydrocarbon wells. We explain how interactions among the parts of such systems can create behavior that is more interesting than individual parts would indicate. Finally, we provide potential directions for future development.

INTRODUCTION

The notion of complex systems has emerged in many contexts over the last several years (Carlson & Doyle, 2002; Ottino, 2003 and references therein). Most, although not all definitions of complexity revolve around the notion of systems whose interconnected parts interact in ways such that the emerging collective behavior appears to the observer non-trivially more interesting than the individual parts would indicate.

It is certainly tempting to try to give a precise definition of what constitutes a complex system in terms of distinguishing system properties. However, we will refrain from doing so. Indeed, such a definition would only be useful to the extent that it could help attract attention to unsolved problems in the systems area and point to directions of future research activity by establishing needs and suggesting opportunities. But that can be accomplished regardless of whether a system is termed complex or not, as long as the research need is well established and the opportunity is clearly suggested. To wit, plant-wide control of large chemical plants is an old and challenging R&D topic that has witnessed significant activity over the last few decades, especially after the introduction of computers in the late 1950s. Yet, to the authors' knowledge, it is only very recently (Åström, 2001) that plant-wide chemical process control has been characterized as "complex", an attribute to which both practitioners and theoreticians of this field would certainly have attested since its early days.

Therefore – partially yielding to the temptation of offering a definition – we will use the attribute "complex" to simply characterize systems which cannot be easily understood (designed, operated, analyzed) using available tools. For such systems there is a need to advance quantitative and qualitative methods for design, operation, and analysis in conjunction with domain-specific know-how. Philosophical and semantic discussions aside, we would claim that such systems are important and pose significant problems which are non-trivial to solve.

In this paper, we are going to focus on our latest work on specific problems related to systems from three different domains: semiconductor manufacturing, hydrocarbon reservoir drilling, and design of dosing regimens for antimicrobial agents. While these domains hardly overlap with each other in terms of the physical systems involved, underlying methods for system study turn out to exhibit many interesting features that are more broadly applicable.

MANUFACTURING OF SEMICONDUCTOR WAFERS

Spatial Uniformity over Semiconductor Wafers

Spatial uniformity over semiconductor wafers is of major importance for high yields in most unit operations of the semiconductor manufacturing industry, such as etching or deposition of thin films and chemical-mechanical planarization. In particular, spatially uniform plasma etching of thin films on semiconductor wafers is a critical step that controls device scaling, circuit performance, magnitude of integration, and overall yield for silicon semiconductors. Roadmaps for future developments in the semiconductor industry call for even larger wafers (450 mm) with even tighter uniformity specifications (SIA, 2004), thus making the spatial uniformity issue even more challenging and important. In a plasma etching reactor, the reactor walls, plasma, and wafer interact in ways that make the analysis of the overall system extremely difficult. To ensure optimal uniformity, the reactor must be appropriately designed, and etch conditions (recipe) must be selected so that uniform etching results. While mathematical models are widely used to design plasma etching reactors, it is practically impossible for such models to be accurate enough to provide quantitative predictions of spatial uniformity of etching. As a result, it is common industrial practice to resort to trial and error for the final configuration of a tool and for recipe development. Therefore, in addition to ultimate accuracy in mathematical model predictions (which is unquestionably desirable) guidance during experimental trial and error is also extremely useful. In particular, methods that untangle the complex relationships among various components of a reactor have clear value for the experimenter, by blending quantitative data with intuition. In that context, the experimenter must be able to:

- Quantify spatial uniformity characteristics, understand similarities and differences between reactors or recipes, and apply criteria for the monitoring of spatial uniformity from tool to tool or run to run, and
- Efficiently optimize spatial uniformity by proper reactor configuration or recipe design.

In the sequel we are going to discuss both of the above two issues in terms of a case study.

Quantitative Characterization of Etching Spatial Uniformity

Because spatial uniformity is usually expressed in terms of a single number (e.g., $3\sigma_{etch\_depth}/\mu_{etch\_depth}$) which must be within specifications, very different spatial profiles of etch rate or depth may result in the same numerical value of uniformity. For example, Fig. 1 shows etch rate profiles for two silicon wafers which correspond to the same numerical uniformity value, but whose corresponding etch patterns are distinctly different (convex versus non-convex), because they were generated under very different experimental conditions. Therefore, no suggestion could be made on how to improve spatial uniformity by simply looking at the effect of such experimental conditions on the numerical value of spatial uniformity. Moreover, a scalar uniformity value hardly suggests any linking between reactor design or etching conditions and observed spatial etching patterns. Consequently, while the use of a single number to describe spatial uniformity is useful for production specification purposes, it masks important information that could otherwise be useful in a number of ways towards tool or recipe performance assessment and improvement.

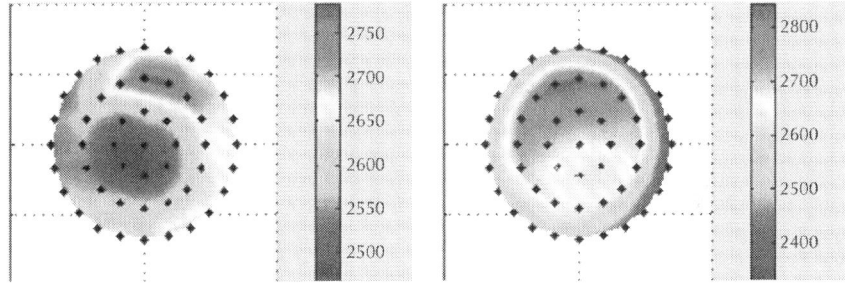

Figure 1: Etch rate profiles on 300-mm wafer surface, interpolated over 49 measurement points. Both wafers correspond to virtually the same numerical uniformity value, but exhibit very different etch patterns.

The difficulties of using a single number to express spatial uniformity are exacerbated when the values of multiple design variables must be selected by experimental trial and error, so that optimal uniformity can result. Things become even more complex when the list of input variables is not a priori clear. As an example, consider a plasma etching reactor with four different design variables, $\mathbf{x} = \{x_1, x_2, W, c\}$, referring to two reactor-specific knobs, power, and reactor configuration, respectively, as discussed in more detail in Section 2.5. The last variable c was not originally in the list of inputs whose effect on uniformity was to be examined, but was rather introduced as a result of processing preliminary experimental data. Nevertheless, with the list **x** known, one could conduct experiments in order to fit a quadratic surface NU = $x^THx + b^Tx + c$ for non-uniformity, NU in the area of the optimum (assuming the latter is vaguely known) and subsequently estimate the optimal settings for **x**. But this would require at the very least (n + l)(n + 2)/2 (i.e. 15 for n = 4) experiments, and preferably more, to mitigate noise effects.

An obvious alternative to a single spatial uniformity value would be to use etch rate information for each point on the wafer surface where the etch rate is measured, as shown for example by the diamond points inFig. 1. Then, standard regression could be used to capture the effect of experimental conditions or reactor design on the spatial uniformity value. However, this alternative would require an inordinately large number of experiments. To avoid this, well-known multivariate analysis methods that take advantage of interdependencies among response variables can be used to greatly reduce the number of experiments needed. Of these methods, principal components regression (PCR) and partial least squares (PLS) are used in this work, because of their simplicity, power, and intuitive interpretation of their results. Specifically, an experimental study on an industrial plasma-etching reactor is presented, to demonstrate how significant improvement of the numerical value of spatial uniformity can result from model-based optimization, where the model is efficiently developed through PCR on scant experimental data.

In the sequel, after presenting a small sample of related literature on spatial uniformity analysis in Section2.3, we present a brief overview of principal component analysis (PCA) and PLS ideas that are relevant to this work in Section 2.4, discuss our experiments and processing of experimental data in Section 2.5, and summarize our conclusions in Section 2.6.

Selected Literature on Spatial Uniformity Analysis

Several authors have applied advanced statistical techniques to understand and explain spatial uniformity. We summarize next a representative sample relevant to this work.

In (Stine, Boning, & Chung, 1997) a variety of techniques such as filtering, spline, and regression-based approaches were used to study wafer-wide spatial variation in integrated circuit processes. Single response surface methodology (RSM) for studying spatial uniformity in semiconductor manufacturing operations has been used in (Guo & Sachs, 1993; Lin & Spanos, 1990; May, Huang, & Spanos, 1991). In (Mozumder & Loewenstein, 1992) response surface models were fitted for uniformity and selectivity and used for optimizing an etch process. The multiple response surface methodology was first used to optimize spatial uniformity in (Guo & Sachs, 1993), where instead of using a single metric, i.e. spatial uniformity, as the fitted variable being predicted by process conditions, the authors chose a few "output characteristics" as the fitted variables. They then applied regression for each of the output characteristics and obtained multiple response surfaces. Guo and Sachs (1993) suggest that multiple response surface models have a number of advantages over single response surface methods, such as robustness to noise, requirement of fewer data, rapid adaptation of models, and compatibility of model forms to process knowledge. Statistical analysis of the multiple response surface methodology has been performed in (Smith, Goodlin, & Boning, 1999). This work also uses multiple response surfaces to optimize uniformity of plasma etching. The output characteristics used in this work are principal component scores, a natural measure of presence of spatial non-uniformity.

The presence of empirical patterns in spatial uniformity of etching profiles is well known in both theory and practice. In (Ha & Sachs, 1999) it is reported that radial uniformity can be controlled by changing certain process variables without affecting circumferential uniformity. This suggests the presence of different features in uniformity patterns and their dependence on different process variables.

Principal component analysis, a well-known statistical technique used in feature extraction (Theodoridis & Koutroumbas, 1999) has been applied to many problems where common features are to be extracted from a series of different 2D patterns such as human faces (Turk & Pentland, 1991). In (Choi, Yoo, & Lee, 2003) PCA is used to monitor process data patterns in a power plant. In (Yu, MacGregor, Haarsma, & Bourg, 2003) PCA is used to characterize and monitor visual features that characterize the quality of the products of a snack food production line. Multivariate analysis is used to analyze spatial features related to surface texture in (Bharati & MacGregor, 1998; Bharati, MacGregor, & Tropper, 2003; Bharati, Liu, & MacGregor, 2004). In (Krischer et al., 1993) the Karhunen–Loeve (KL) expansion is used to extract dominant spatial structures from spatio-temporal data of CO oxidation on Pt surface. In (Rigopoulos & Arkun, 1996) the KL expansion is used to extract a small number of dominant modes the describe spatial uniformity of sheets produced in a paper manufacturing machine. Application of PCA to extract common features in etch rate data over a wafer was reported in (Nikolaou & Bailey III, 2002). PCA has also been used in (White, Boning, Butler, & Barna, 1997) to do an indirect in situ spatial characterization of wafer state using optical emission spectroscopy data.

Latent-Variable Methods for Spatial Uniformity

Latent-variable methods are multivariate techniques, which starting with a vector of variables generate a vector of new (latent) variables (also called modes or components) such that: (a) the components are a linear – or sometimes non-linear – combination of the original variables and (b) a "small" number of components is enough to capture a property of the original variables, such as total variance (principal component analysis), correlation (factor analysis), differences among the original random variables (discriminant analysis), explanation of the correlation between the original variables and response variables (canonical correlation analysis), or explanation of the variance of response variables (partial least squares). Of these methods, PCA and PLS are most relevant for the uniformity problem we study. These methods are well discussed in literature (e.g., Jackson, 1991, Jolliffe, 1986 and Wold, 1982) and we are not going to discuss them in detail here. A brief description of PCA and PLS aspects that are relevant to wafer image processing is given next.

Principal Components Analysis

Principal component analysis (PCA) is also known as the discrete Karhunen–Loeve transform in image processing and pattern recognition (Theodoridis & Koutroumbas, 1999) and feature extraction applications (Savoji & Burge, 1985), and proper orthogonal decomposition in the study of dynamical systems (Holmes, Lumley, & Berkooz, 1996). PCA is usually presented in a statistical context, but that is not necessary. As discussed below, an approximation context is enough, although a statistical interpretation is undoubtedly very useful. PCA starts with a data matrix where each column **x**i contains m values of a single variable xi and each row corresponds to a sample of all n variables {x_1, ..., xn} and approximates it by a matrix $\hat{\Xi}$ of lower rank after finding.

$$\mathbf{X} \hat{=} [\mathbf{x}_1 | \cdots | \mathbf{x}_n] \in \Re^{m \times n} \tag{1}$$

where each column xi contains m values of a single variable xi and each row corresponds to a sample of all n variables {x_1, ..., x_n} and approximates it by a matrix $\hat{\Xi}$ of lower rank after finding

$$\min_{\mathrm{rank}(\Xi) = r < \min(m,n)} \|\mathbf{X} - \Xi\|^2 \tag{2}$$

It can be shown (Dewilde & Deprettere, 1988) that if the matrix norm in Eq. (2) is either the induced 2-norm[1] or the Frobenius norm,[2] the optimal solution is

$$\hat{\Xi} = \sum_{j=1}^{r} \underbrace{\sigma_j \mathbf{u}_j \mathbf{v}_j^T}_{\mathbf{y}_j} = \sum_{j=1}^{r} \mathbf{y}_j \mathbf{v}_j^T \qquad (3)$$

where $\sigma_1 \geq ... \geq \sigma r$ are the r largest singular values of $X = \sum_{i=1}^{n} \sigma i u v_i^T = USV^T$, and $\mathbf{u}_i$, $\mathbf{v}_i$ are the corresponding standardized singular vectors. In addition, $\left\| X - \hat{\Xi} \right\|_{i2}^2 = \sigma^2 r + 1$ and $\left\| X - \hat{\Xi} \right\|_{i2}^2 = \sigma^2 k = r + 1 \sigma_k^2$.

$$y_j \hat{=} \sigma_j \mathbf{u}_j = \mathbf{X} \mathbf{v}_j \qquad (4)$$

The vectors which are orthonormal to each other, and $\mathbf{v}_j$ are usually called the scores and loadings, respectively. Each scores vector, $\mathbf{y}_j$, can be shown to contain the values of a new variable yj = [x_1 ..., xn]$\mathbf{v}_j \equiv \mathbf{x}^T \mathbf{v}_j$, also called a component. Because $\|y_j\|^2 = \|\sigma_j u_j\|^2 = \sigma^2$; by Eq. (4), the magnitude $\sum_{j=1}^{r} \|y_j\|^2 = \sum_{j=1}^{r} \sigma_j^2$ of the first r (principal) components yj captures most of the total magnitude $\sum_{i=1}^{n} \|xi\|^2 \hat{=} \|x\|_F^2 = \sum_{all} j\sigma_i^2$ in the original data. When x and y are random variables, magnitude corresponds to variability.

Principal Components and Wafer Imaging

The loading vectors $\mathbf{v}_j$, can be shown to be interpretable in two ways, which provide good intuition for the description of wafer-wide uniformity:

- Each loading vector $\mathbf{v}_j$ can be thought of as the vector of weights v_{ji}, $i = 1, ..., n$ used to construct the component y_j as a linear combination of the original variables x_i, $i = 1, ..., n$, i.e.

$$y_j = \sum_{i=1}^{n} v_{ji} x_i.$$

(5)

Thus, the importance of each x_i for each y_j can be assessed by the corresponding magnitude of v_{ji}. An example of the interpretation of loadings as weights is shown in Fig. 3.

- Each loading vector v_j, can also be thought of as one of the basic shapes whose scores-weighted linear combination produces each observed shape (row $[x_{1k}, ..., x_{nk}]$ of $\mathbf{X}$) contained in the data, as

$$[x_{1k} \cdots x_{nk}] = \sum_{j=1}^{r} y_{kj} \mathbf{v}_j^T.$$

(6)

Such basic shapes frequently have physical significance that can be easily recognized by practitioners and can be intuitively associated with experimental factors. An example of the interpretation of loadings as basic shapes is shown in Fig. 2.

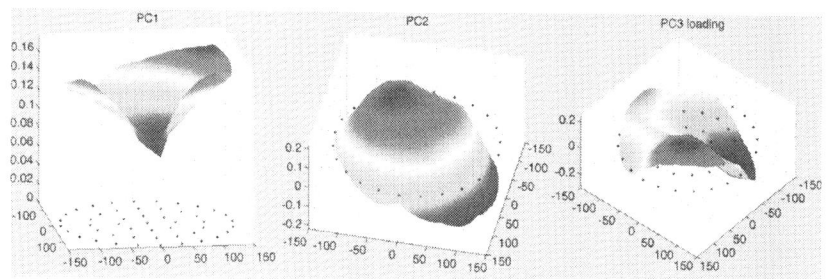

Figure 2: Top and angle views of loadings as basis shapes (contour surfaces) corresponding to the first 3 principal components for the same set of wafers as in Fig. 3 (Nikolaou & Bailey III, 2002).

There are many criteria for selecting the number r of principal components that capture "enough" of the variability of the original data. We have found the following two most useful for the description of wafer-wide etch patterns (Fig. 3):

Reconstruction of the original data from loadings and scores (Eq. (5)) retains all qualitatively features that are important (Fig. 4).

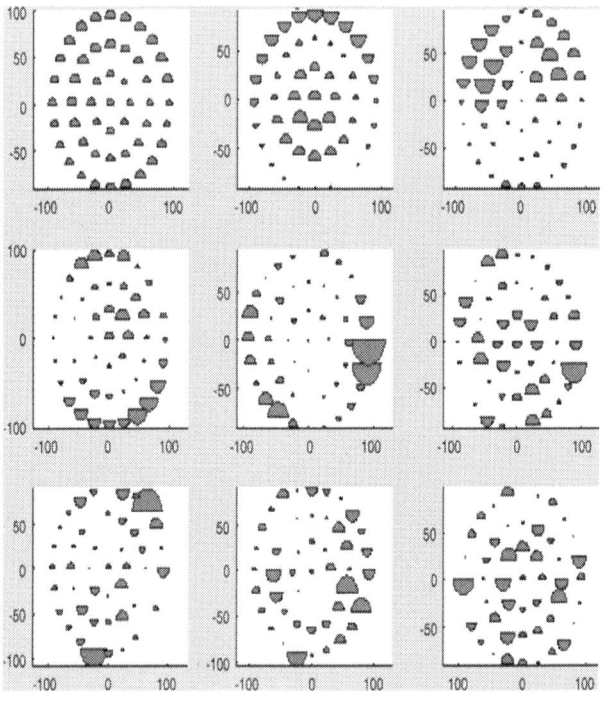

Figure 3: Vectors of loadings for experimental data collected from 9 etched wafers, on each of which etch rate measurements were taken over 49 points (Nikolaou & Bailey III, 2002). The area of each semi-disk corresponds to the magnitude of the weight and the orientation to the weight's sign. Note that weights are fairly smoothly patterned for the first few loadings (corresponding to principal components) in contrast to the remaining loadings (corresponding to non-principal components), which appear erratic due to noise.

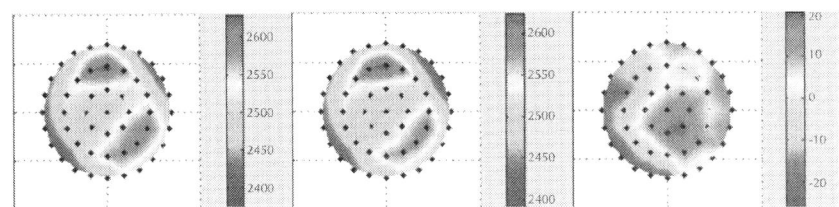

Figure 4: Original etch rate data, data reconstructed by using 3 principal components, and residual errors for one of the 9 wafers referred to in Fig. 3 (Nikolaou & Bailey III, 2002).

The cross-validation error is minimized.

Regression with Principal Components

Principal components and Eq. (3) can be used effectively to reduce dimensionality in multivariate regression problems, as where $M \varepsilon \chi^{m \times e}$ is the data matrix; $\Theta \in \Re^{\ell \times n}$ and $\Theta c \in \Re^{\ell \times r}$ are the original and compressed parameter matrices, respectively; and $Y_c \varepsilon \chi^{m \times r}$ is the scores data matrix, containing values for the principal components $\{yi\}_{i=1}^{r}$, which are (much) fewer than the original variables $\{xi\}_{i=1}^{n}$, thus allowing regression to be performed using scant experimental data. We demonstrate this in the sequel.

$$\min_{(\cdot)} \lVert \mathbf{X} - \mathbf{M}\Theta \rVert_F$$
$$= \min_{(\cdot)} \lVert \mathbf{USV}^T - \mathbf{M}\Theta \rVert_F \approx \min_{(\cdot)} \lVert \underbrace{\mathbf{U}_c \mathbf{S}_c}_{Y_c} \mathbf{V}_c^T - \mathbf{M}\Theta \rVert_F$$
$$= \min_{(\cdot)} \lVert \mathbf{Y}_c^T - \underbrace{M\Theta \mathbf{V}_c}_{\Theta_c} \rVert_F \qquad (7)$$

Partial Least Squares

While PCA has been fairly well developed and understood since its inception in the early part of the 20th century (Jolliffe, 1986), PLS was developed much later (Wold, Ruhe, Wold, & Dunn, 1984).

In his historical review of PLS, Geladi (1988) cites the inventor of PLS (Wold, 1982) as declaring "1977 as the birthdate of PLS". While there has been significant practical experience with PLS, it is only recently that its properties have been rigorously understood (Breiman & Friedman, 1997; Butler & Denham, 2000; Frank & Friedman, 1993; Garthwaite, 1994 and Goutis, 1996; Helland, 1990 and Höskuldsson, 1988). The PLS algorithm (Geladi & Kowalski, 1986) works with input and output data matrices **X** and **Y**. After a first pair of scalar latent input and output variables is constructed as linear combinations of input and output variables, respectively, that are maximally correlated, a linear (or non-linear: Geladi, Martens, Hadjiiski, & Hopke, 1996; Holcomb & Morari, 1992; Malthouse, Tamhane, & Mah, 1997; McIntosh, Bookstein, Haxby, & Grady, 1996; Qin & McAvoy, 1992; Wold, 1992) model is fitted between these latent input and output variables. The part of the data that can be parametrized in terms of the latent input and the latent output that is predicted by the fitted model is subtracted from the original data **X** and **Y**. The process is repeated on the remainders, until all possible latent variables are constructed.

Case Study

Experimental Setting

All experiments were conducted at the facilities of Lam Research Corporation in Fremont, CA. A blanket 20 kÅ film of silicon oxide and 3 kÅ film of silicon nitride on wafers of 200 mm diameter were etched in an inductively coupled plasma-etching reactor. Etch rates were inferred at 49 points on each wafer, as shown in Fig. 1, by measuring the thickness of oxide or nitride films before and after etching. Film thickness for all oxides was measured using OptiProbe 1000, an ellipsometry-based metrology tool (ThermaWave). Nitride film thickness was measured using ASET-F5x, a thin-film metrology tool (KLA-Tencor). All computations were done using Matlab (Mathworks). For some PLS computations the PLS Toolbox

(Eigenvector Research) was used. Details can be found in (Misra, 1993).

As mentioned in Section 2.2, the objective of this study was to study the effect on spatial etching uniformity of four design variables: two process knobs, called x_1 and x_2, bias power W, and a reactor configuration parameter, c. As already explained, one could fit a quadratic surface of the form NU = $x^THx + b^Tx + c$ to experimental data, and produce the next estimate for optimal settings at x = (–1/2) $H^{-1}b$. But this would require at the very least (n + l)(n + 2)/2 = 15 (for n = 4) experiments, and preferably more, to mitigate noise effects. In addition, it would provide very little intuition regarding the connection between observed etching patterns and the four input variables. Finally, it would make relatively little use of qualitative prior experience with this kind of reactor.

To address these issues, we applied the multivariate analysis ideas presented in the preceding sections. We are discussing below how these ideas helped guide the experiments and provide intuition regarding the causes of etching non-uniformity.

Effect of (X_1, X_2) on Etching Uniformity of Silicon Oxide Wafers

Preliminary Experiments

At the beginning of the study, engineers who had used reactors similar to the one used in this study pointed out that the two input variables x_1 and x_2 were known to have a significant effect on etching uniformity. To verify this, we conducted four preliminary experiments (oxide wafers #1–4 in Table 1). All process conditions were kept constant, except the two knobs x_1 and x_2 whose values were changed using the standard response surface methodology (e.g., Montgomery, 2001).

Table 1: Experimental values of $x_1 x_2$ and NU

Wafer #	x_1	x_2	NU = $3\sigma/\mu$ (%)
1	1.00	−1.00	5.4
2	−0.95	−1.00	11.7
3	1.00	1.00	5.1
4	−1.00	0.98	12.4
5	0.64	0.030	6.2
6	1.00	0.0016	5.7
7	−0.99	0.0094	12.3
8	0.006	1.00	8.2
9	2.02	−0.90	3.4
10	4.61	4.98	12.3
11	3.74	3.02	8.1
12	6.17	4.99	24.6
13	0.13	−3.00	7.3
14	2.79	1.64	5.6
15	2.80	−1.67	6.2
16	−0.96	−5.00	4.9
17	1.67	0.0091	3.9
18	2.02	−0.90	3.1

Direct Model: Effect of (X_1, X_2) on Non-Uniformity Nu

The first four experiments in Table 1 cannot be used to directly determine how x_1 and x_2 affect NU, let alone determine the optimum of NU, if a quardratic-surface model of the formis used. At best, one can say that the value of a_1 in Eq. (8) is not needed in order to determine the optimum of NU. Reduced-rank regression on Eq. (8) would not help either. For example, an attempt to use PLS regression to fit the first four data points by a model of the form Eq. (8), yielded the convex quadratic function

$$NU = a_1 + a_2x_1 + a_3x_2 + a_4x_1x_2 + a_5x_1^2 + a_6x_2^2 \qquad (8)$$

$$NU\% = 3.02 - 3.55x_1 + 0.0242x_2 + 2.85x_1^2$$
$$+ 2.95_2^2 - 0.176x_1x_2. \qquad (9)$$

The resulting optimal values for x_1 and x_2 were used to etch a fifth oxide wafer (Table 1). As expected, the experimentally observed uniformity was not optimal. In fact, it was worse than that observed in experiments 1 and 3. Thus, a model that maps the values of x_1 and x_2 to the value of 3σ non-uniformity cannot be used to determine optimal values of x_1 and x_2 from just four experimental points.

Indirect Model: Effect of {X_1, X_2} on Etch Rate Profile

Scaled etch rate data from the first four oxide film wafers were put in a 4 × 49 matrix **X**, c_i. Eq. (1), to perform PCA. Each column of this matrix represents scaled measurements at a particular wafer point, and each row represents scaled wafer-wide measurements on one wafer. Scaling was done by removal of the wafer-wide average from measurements for each wafer. The important issue of scaling for PCA is discussed in more detail in (Misra, 1993).

PCA indicated that three main independent shapes (Fig. 5) in the non-uniformity patterns (out of a maximum of four for four data points) can capture 97.6, 99.9 and 100% of variability in the **X** data. In fact, almost all variability (>97%) is captured by just the first principal component, suggesting that removal of that component would result in substantial improvement of spatial uniformity. Therefore, the effect of x_1 and x_2 on NU was examined. Linear regression on the first principal component yielded with R^2 = 0.994 and F = 88.6 at 95% confidence, indicating high linear dependence on x_1 and x_2. Consequently, by appropriate selection of x_1 and x_2, the non-uniform shape represented by the first principal component, which is independent of the other principal component shapes, can be removed, with concomitant uniformity improvement. According to Eq. (10), an appropriate selection for

x_1 and x_2 is $(x_1, x_2) = (2.43, 0)$, given that $x_2 = 0$ is a preferred value for x_2 when possible, based on physical arguments. No experiment was performed at that value of (x_1, x_2), but, as shown in the sequel, this value is very close to the optimum as determined by additional experiments, resulting in predicted (interpolated) non-uniformity value below 5%. This is an improvement over previous values, and makes the case for the value of using PCA to optimize uniformity.

$$PC1_{score} = 423 - 174x_1 + 9.79x_2$$

(10)

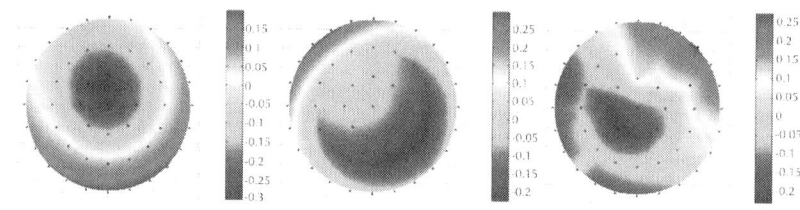

Figure 5: Contour plots of loadings corresponding to PC1, PC2 and PC3 for oxide wafers, calculated from the first four experiments in Table 1.

Additional Experiments for the Effect of (X_1, X_2) on Oxide Wafers

To verify the insight provided by the preceding principal component analysis, another three experiments were conducted on oxide wafers #6–8, as shown in Table 1. The intermediate value of 0 was considered for x_1 and x_2. The first eight values of NU in Table 1 were used to fit the quadratic model (cf. Eq. (8)),

$$NU\% = 8.14 - 3.391x_1 + 0.134x_2 + 0.753x_1^2 \\ - 0.212x_2^2 - 0.250x_1x_2,$$

(11)

The above model is not convex, but has a saddle point, indicating that it does not represent the effect of $\{x_1, x_2\}$ on NU well. In fact, the model indicates strong dependence of NU on x_1 and weak dependence on x_2, as shown in Fig. 6a. Indeed, while the optimal value of x_1 appears to be slightly above 2, the optimal value of x_2 cannot be determined.

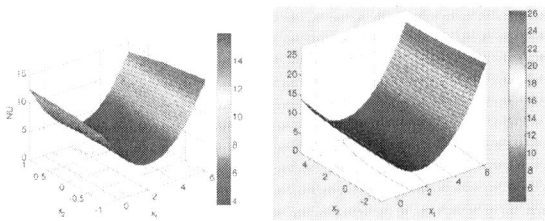

Figure 6: Non-uniformity surface fit according to Eq. (11) for: (a) the first eight experimental data points and (b) all data points shown in Table 1. Note the strong dependence of non-uniformity on x_1 as contrasted to weak dependence of non-uniformity on x_2.

To address this issue, we performed PCA on the first eight experimental data points (Table 1). The resulting counterpart of Eq. (10) is with $R^2 = 0.993$ and $F = 358$ at 95% confidence. For x_1 to be slightly above 2, as required above, the value of x_2 must be negative. Because x_2 must be as close to 0 as possible (due to reactor operability issues), the lower bound on x_2 used in the prior experiments was targeted, i.e. −1. The actual conditions implemented on the ninth experiment were (2.0, −0.90) [3] on wafer #9 (Table 1). The resulting value of NU was 3.4%, a clear improvement over previous uniformity values.

$$PC1\,\text{score} = 426 - 171x_1 + 7.67x_2 \tag{12}$$

Fig. 7 shows the measured etch rate pattern for experiment #9 in Table 1. Note that at these conditions Eq.(12) suggests that the first principal component has been practically removed. Note also that the etch rate pattern at such conditions looks almost like the second principal component in Fig. 8.

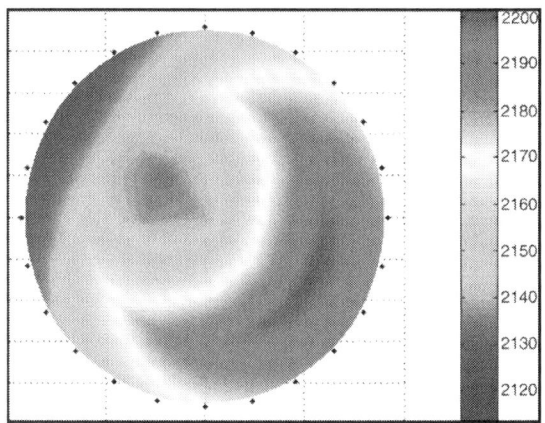

Figure 7: Etch rate pattern for oxide wafer with PC1 almost zero. (Note the sign reversal in comparison to PC2 in Fig. 8).

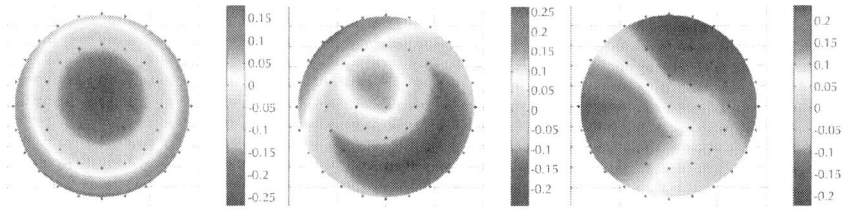

Figure 8: Contour plots of loadings corresponding to PC1, PC2 and PC3 for oxide wafers, calculated from all experiments in Table 1. Compare to Fig. 5.

Full Set of Experiments for the Effect of (X_1, X_2) on Oxide Wafers

To cover a wide operating area and assess the validity of predictions made about optimal uniformity from scant experimental data, another eight experiments were conducted on oxide wafers #10–17 (Table 1). Finally, a last experiment was performed to verify by repetition the results of experiment #9. The non-uniformity results from all 18 experiments can be fit by the convex quadratic surface shown in Fig. 6b. The optimal values suggested by Eq. (13) are (x_1, x_2) = (1.76, 0.30) with predicted non-uniformity 4.4%. Given that this value is higher than the experimentally observed values of slightly above 3% (wafers #9 and 18 in Table 1), one can infer that the non-uniformity values for the optimal settings of (x_1, x_2) suggested after four and eight experiments are most probably between 3 and 4%, which makes these suggestions even stronger.

$$NU\% = 7.63 - 3.68x_1 + 0.0964x_2 + 1.05x_1^2$$
$$+ 0.0286x_2^2 - 0.0645x_1x_2 \quad (13)$$

PCA on all 18 experimental data (18 × 49 matrix **X**) produces loadings for the first three principal components shown in Fig. 8. The first three principal components capture 85, 98.9, and 99.9% of variability, respectively. Analysis by the PRESS statistic (Fig. 9) indicates that the cross-validation error becomes minimum when nine PCs are retained, but three PCs are almost as good.

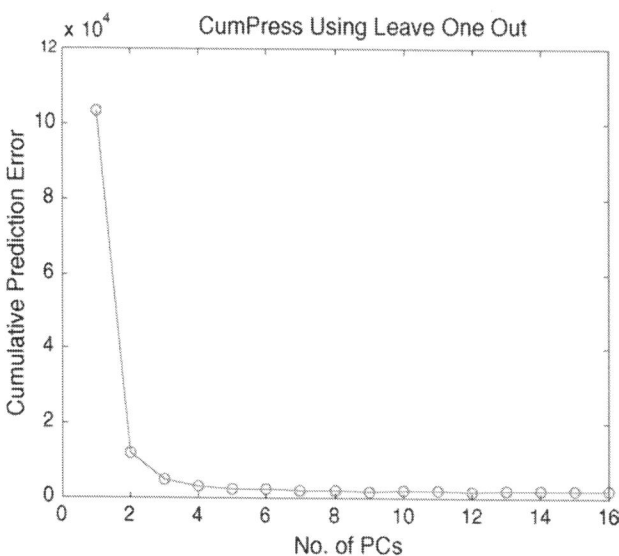

Figure 9: PRESS for PCA using all 18 experimental data points in Table 1.

Also, addition of some noise to the data changes the loading of the fourth PC considerably, but has little to no effect on first three PCs, as seen in Fig. 10, corroborating the choice of up to three components.

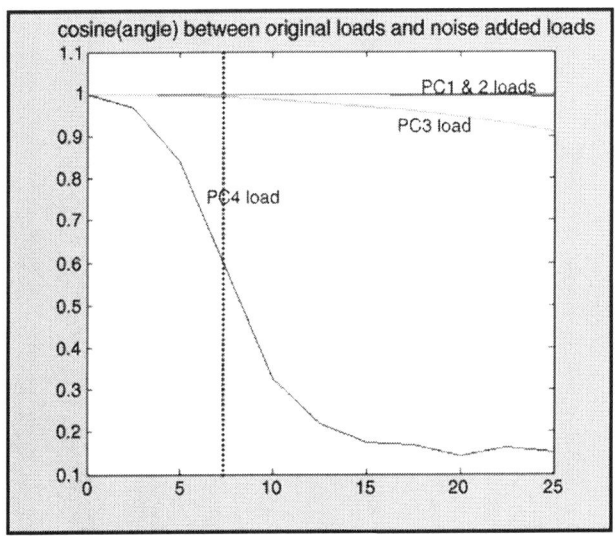

Figure 10: Change in loadings with addition of noise for oxide wafers. The vertical line shows the amount of noise expected in this system, according to the experiments.

Note that the trends captured by the shapes of the loadings in Fig. 8 are qualitatively similar and quantitative quite close to those in Fig. 5, which were generated by merely four experimental points. This suggests that these shapes are strong indicators of the intrinsic behavior of the system at hand, and that PCA can easily reveal that behavior.

$$\text{PC1 score} = 389 - 234 x_1 - 5.55 x_2 \tag{14}$$

As before, a linear regression model can be developed for all 18 points as with $R^2 = 0.969$ and F = 200 at 95% confidence level.

This verifies that the shape present in the etch profile due to the first principal component, i.e. the concentric ring pattern shown in Fig. 5 and Fig. 8 can be removed, and uniformity can be improved, by adjusting the two process knobs x_1 and x_2.

Compared to the first principal component, the score values corresponding to the second principal component (Fig. 11) are not linearly dependent on the two process knobs x_1 and x_2. Linear regression in this case resulted in $R^2 = 0.63$ and $F = 11.3$ at 95% confidence level.

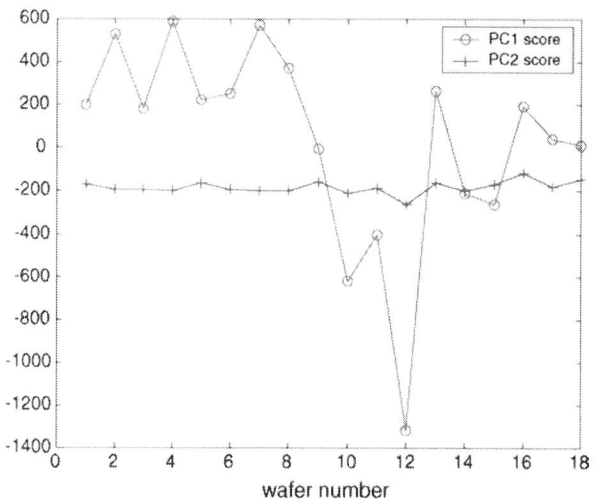

Figure 11: Scores for PC1 and PC2 for oxide wafers.

The score values for the third principal component also appear not to be linearly dependent on the two process knobs x_1 and x_2. ($R^2 = 0.53$ and $F = 8.9$ at 95% confidence level).

This means that uniformity can be controlled by changing the score values corresponding to the first principal component, which in turn can be controlled by changing the process knobs. However, it is not possible to produce any desired shape, since there are three principal shapes and by adjusting the knobs (x_1, x_2) only the first principal shape can be removed. The other two shapes will remain in the wafer.

Effect of (X_1, X_2) on Etching Uniformity of Silicon Nitride Wafers

To determine how basic features that are present in etch patterns vary with wafer film type, etching of silicon nitride film wafers was also studied. For nitride wafers etching, the main etch step recipe conditions were kept identical to the conditions used for oxide wafers. The same process knobs x_1 and x_2 were also varied as in the etching of oxide wafers (Table 2). In total, 11 wafers were etched. The first wafer was etched at the point where the best oxide uniformity was seen (point #9 in Table 1). The remaining points are from a standard 2k design around the four corners, at (0, 0) and at (0.75, 0) in the four quadrants. The resulting non-uniformity values follow the pattern shown in Fig. 12.

Table 2: Experimental values of (x_1, x_2) at which nitride etching was done

x_1	x_2
−2.23	−0.90
0.99	1.01
0.97	−1.02
−1.00	0.99
−1.02	−1.02
0	−0.01
−0.48	1.02
0.76	0.011
0	0.74
−0.76	−0.019
0	−0.0089

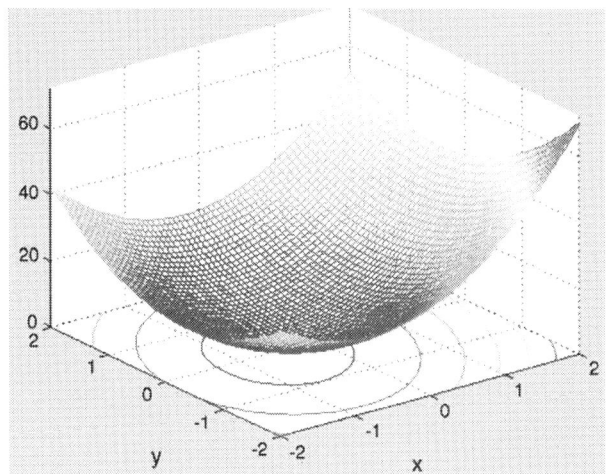

Figure 12: Non-uniformity surface fit according to Eq. (8) for all 11 experimental data points shown in Table 2. Note the approximately equal dependence of non-uniformity on both x_1 and x_2 (cf. Fig. 6).

Fig. 13 shows the contour plots for loadings corresponding to PC1, PC2, and PC3, respectively. It can be observed that the loading corresponding to PC1, whose score changes with changing the process knobs, is similar to the loading corresponding to PC1 in the case of oxide etching (Fig. 8). Consequently, PC1 is process-knob dependent and does not change much for these two film types. However, the shapes of the loadings corresponding to PC2 and PC3 for nitride etching are very different from those for oxide etching.

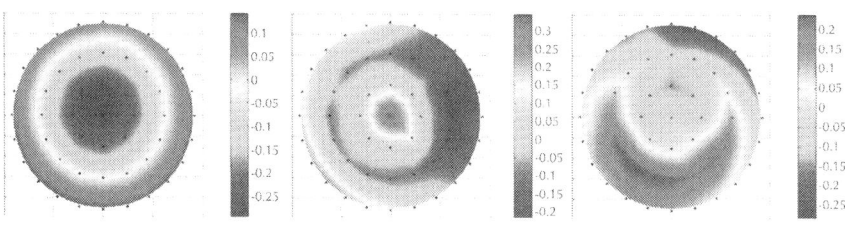

Figure 13: Contour plots of loadings corresponding to PC1, PC2, and PC3 for nitride wafers. Compare to Fig. 8.

Linear regression yields with $R^2 = 0.934$ and $F = 63.3$ at 95% confidence level. Linear regression between PC2 score and the two process knobs results in $R^2 = 0.41$ and $F = 3.13$ at 95% confidence level. The regression fit for PC2 is much worse than for PC1, as was the case for oxide etching.

$$\text{PC1 score} = -88.8 - 268x_1 - 65.8x_2 \tag{15}$$

Summary of Effect of (X_1, X_2) on Oxide and Nitride Wafers

For both oxide and nitride wafers, a basic shape (co-centric rings of PC1 loading) in etch patterns is linearly related to (x_1, x_2), while all other shapes (PC2 loading and above) are not clearly affected by (x_1, x_2). Only few experiments are needed to get the relation between the PC1 score and (x_1, x_2), and then the process condition at which the PC1 score becomes zero and uniformity is improved can easily be characterized.

Effect of Bias Power W on Etching Uniformity of Oxide Wafers

In the previous sections it was seen that the PC2 score cannot be removed by adjusting the process knobs (x_1, x_2), and that the shape of loading corresponding to PC2 is dependent on the kind of wafer being etched (cf. Fig. 8 for oxide versus Fig. 13 for nitride). Therefore, non-uniformity contributed by PC2 cannot be removed by adjusting the knobs (x_1, x_2). However, it is known from experience that changing the bias power while keeping other conditions constant in an inductively coupled plasma-etching reactor affects non-uniformity. Therefore, it was decided to study the effect of bias power on PC scores, and PC2 in particular.

To verify this knowledge by covering a broad range of conditions, six additional oxide wafers (included inFig. 14) were etched at two additional bias power conditions (900 and 700 W) and at three different values of (x_1, x_2), and wafer-wide etch rates were measured. It was assumed, and verified by analysis of residuals, that the loadings developed in Section 2.5.2 for the effect of $\{x_1, x_2\}$ on uniformity at fixed bias power (1100 W) remain the same when different bias power values are used. This is also supported by comparison between Fig. 8 and the second column in Fig. 18. Consequently, the values of the PC1, PC2 and PC3 scores for the variable power case are calculated, according to Eq. (4), by multiplying the scaled etch rates by the loadings corresponding to PC1, PC2 and PC3 from the constant bias case. The results show that bias power has a strong effect on the PC2 score, namely the higher the bias power, the lower the PC2 score, as shown in Fig. 14.

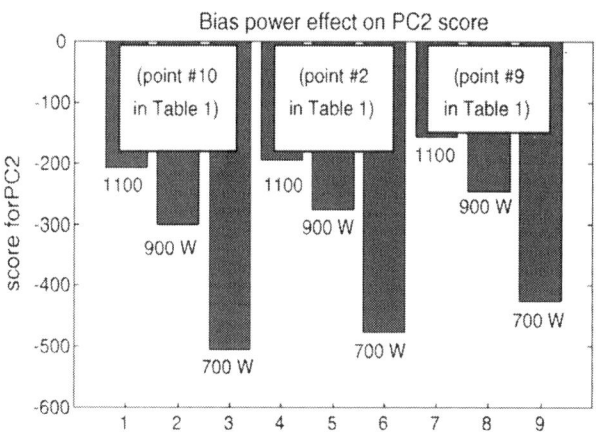

Figure 14: PC2 scores for different bias powers.

Consequently, by proper setting of x_1, x_2, and bias power, the contributions of the first two basic shapes to the overall etch pattern can be made very small and overall non-uniformity can thus be decreased. This was verified experimentally by etching at 1300 and 1400 W. Fig. 15 shows the values of PC2 score for the same values of x_1, x_2. It is seen that the PC2 score does not vary much when bias power is very high and the minimum for this recipe with oxide wafers is close to –80. At such settings, i.e. bias power of 1300

W, and with proper selection of $\{x_1, x_2\} = (3.25, -1.67)$ to give low PC1 score, a non-uniformity value of 1.9% was achieved. By comparison, the lowest non-uniformity value for this recipe in this tool for an oxide wafer, known by trial and error, was 3.4%. The etch rate pattern at that setting is shown in Fig. 16.

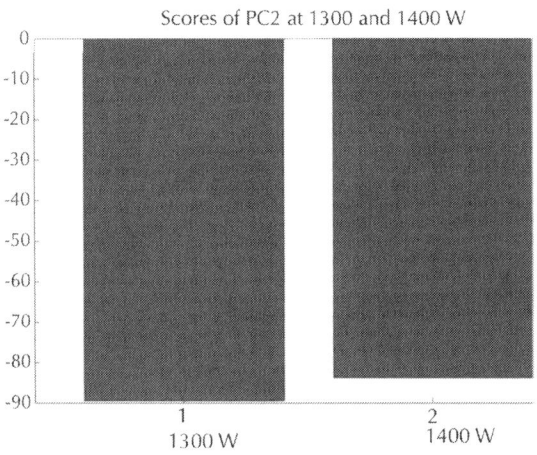

Figure 15: PC2 scores at high bias powers.

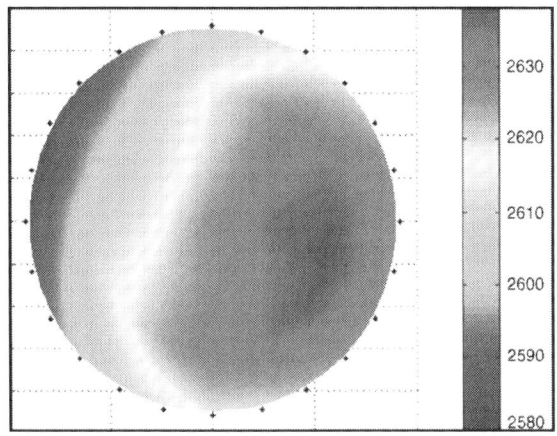

Figure 16: Etch rate pattern for oxide wafer corresponding to non-uniformity value of 1.9%.

Effect of Reactor Configuration on Etching Uniformity of Oxide Wafers

The preceding PCA on etch rate data with constant bias power for oxide wafers showed that there are three principal shapes which contribute to the non-uniformity of etch patterns. These three shapes have different amounts of contributions for different wafers and the variation in contribution is highest for the shape corresponding to PC1. This contribution varies with the settings of process knobs (x_1, x_2).

It was also observed that the PC2 score remains almost constant for different wafers and that the shape corresponding to PC2 bears some resemblance to a known hardware asymmetry in the plasma reactor. It was hypothesized that by removing that hardware asymmetry, uniformity could be improved by making the PC2 score smaller in magnitude. While it was not feasible to make major hardware changes that would completely remove that asymmetry, it was possible to make a few minor changes that could decrease, though not entirely remove that particular asymmetry.

Experiments were done after making each of such minor changes in the hardware. For all of these experiments all recipe conditions were kept fixed. The process knobs x_1 and x_2 were kept at the same values as in experiment #9 of Table 1, for which the unmodified reactor produced a 3σ non-uniformity value of 3.4%. In this case too the PCs produced by the original oxide wafer data (Fig. 8) were used to get the PCA model, and the etch rate data from the new experiments were projected on the model to get the PC scores or the contribution of various shapes. The main interest in this case was in the PC2 score and seeing if it changes by changing the hardware configuration.

Fig. 17 shows the effect of making hardware changes on the PC2 score. It can be seen that the values of the PC2 score are less than those observed for the unmodified reactor where the PC2 score was close to −200. For the best modification achieved with minor changes a PC2 score of −130 is observed. For that particular modification, the 3σ non-uniformity was experimentally observed to be 2.5%, an improvement over the 3.4% value that was obtained without hardware modifications.

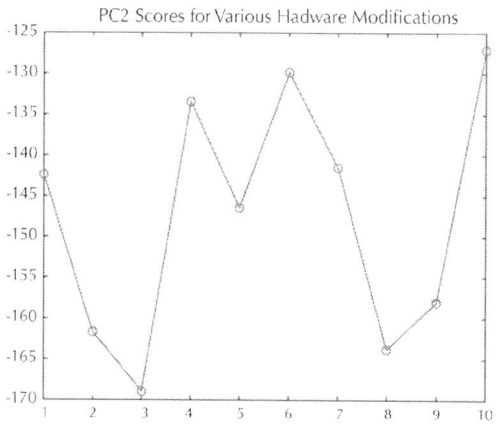

Figure 17: Effect of hardware modification on PC2 score for oxide wafers with (x_1, x_2) kept constant at the condition that gives 3.4% non-uniformity in the unmodified reactor.

Application of PLS To Feature Extraction and Relation to Process Variables

In the preceding sections, it was shown that etch rate data from oxide wafers could be expressed as a linear combination of three basic shapes (Fig. 8). These shapes were obtained using PCA and indicated "directions" of maximum variability in the etch rate data. The extent to which these patterns were present in the etch rate data of any given wafer was given by principal component scores. The scores on PC1 and PC2 were related to process variables via an ordinary least squares regression.

Even though PCR produced quite satisfactory results, we also performed PLS on the same data, for comparison purposes. We wanted to determine the relationship between features extracted by PCA (loadings) and PLS (latent outputs). The results are shown in Fig. 18. It can be seen that the first two features (latent variables) are similar to the first two PC features. This was to be expected as we have seen that changing the process variables can control the first two PC scores. The third PC was independent of the process variables and it can be seen that the third feature that PLS finds is different from the third PCA feature. The third PLS feature could be due to noise because two latent variables in the PLS model capture 80.81% variation in etch rate data and the three latent variables capture 81.68% variation, i.e. the third latent variable does not provide much additional information.

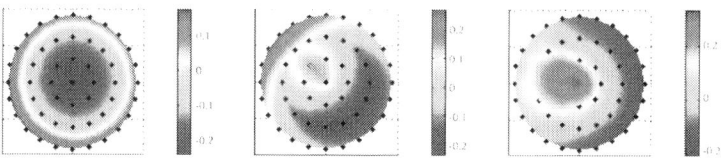

Figure 18: Features extracted by PLS (latent outputs) from all etch rate data for oxide wafers (Table 1). Compare to PCA features in Fig. 8.

Fig. 19 shows latent variables (LV) plotted against PC scores for PC1 and PC2. While LV1 has a perfect linear relationship with PC1 score, LV2 does not have such a strong relationship with PC2. This is also expected as the PC2 shape could be controlled by bias power, but there was not a strong linear dependence of the LV1 score on bias power. However, PC1 score had very strong linear correlation with (x_1, x_2) and thus PC1 and LV1 are almost identical.

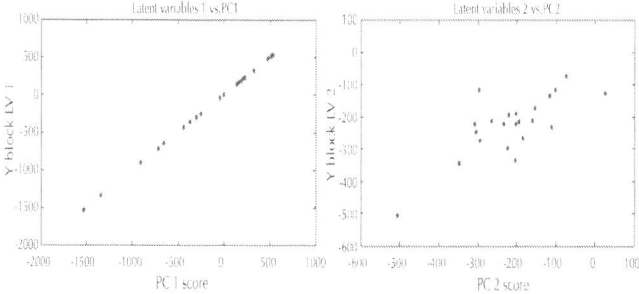

Figure 19: Output latent variables compared with principal components.

Summary: Complexity in Plasma Etching Uniformity

The complexity of plasma etching reactors comes from the interaction among plasma, wafer, and reactor. The main goal of this work was to demonstrate how this complexity can be reduced by combining experiments and analysis of experimental results. In particular, we showed how to reduce asymmetrical etch rate patterns by adjusting process and equipment conditions. It was claimed, and experimentally shown on a commercial plasma-etching reactor, that decomposition of non-uniformity patterns into a small number of uncorrelated shapes and subsequent building of a regression model that maps process and equipment conditions to such shapes can indeed be used for this purpose. The experiments conducted in this work revealed salient non-uniformity shapes and connections to equipment features which were sensible, after the fact, but had not been previously connected with uniformity on the basis of intuition alone. Other studies show that in other etch tools too there are two to three salient shapes as well (Yadav, 2002). It was found that each of the salient non-uniformity shapes can be removed (thus resulting in overall non-uniformity reduction) almost independently of the remaining shapes by appropriate selection of a process or equipment variable. In addition to efficiency, decomposition of non-uniformity patterns to uncorrelated non-uniformity shapes provided valuable insight into the causes of non-uniformity, thus substantially reducing the complexity of the system.

The study points to issues that warrant further investigation, such as:

- Treatment of significant non-linearities.
- Efficient design of experiments when various non-trivial constraints are present (e.g. minimum etch rate).
- Connection between experimental results and first-principles plasma etching simulators.

COMBATING MICROBIAL RESISTANCE TO ANTIMICROBIAL AGENTS

The Need to Combat Microbial Resistance and the Role of Modeling

Microbial resistance to antimicrobial agents (e.g. resistance of pathogenic bacteria to antibiotics) is a rapidly spreading problem with potentially grave consequences (Gold & Moellering, 1996; Morens, Folkers, & Fauci, 2004; Neu, 1992; Levy, 1998 and Varaldo, 2002). Repeated warnings have been issued of microbial wars (Koshland, 1992), new plagues (Koshland, 1992), worldwide calamities (Kunin, 1993), and new apocalypses (Ash, 1996 and Levy, 1994). Many clinicians are concerned that common infections may be untreatable in the near future (Cohen, 1992 and Drlica, 2001). The total cost of antimicrobial resistance to U.S. society is nearly $5 billion annually (National Institute of Allergy and Infectious Diseases, 2004). Broad-spectrum antimicrobial resistance in HIV, tuberculosis, Gram-negative bacteria (e.g. Pseudomonas aeruginosa) and agents implicated in bioterrorism (e.g. anthrax) are especially problematic and have world-wide implications. There is an urgent need for effective new antimicrobial agents, but the rate of their development has not kept pace with the increase in resistant pathogens (Wise, 1998). Therefore, it is imperative to both: (a) develop methods that accelerate the development of new antimicrobial agents and (b) preserve the efficacy of available agents through judicious use. Systematic efforts towards these goals include attempts to understand how to model the effect of an agent on a microbial population that is not homogeneous but consists of microbes of varying susceptibility (or resistance) to the agent. The ultimate objective of such modeling efforts is to develop methods for the design of agent dosing regimens that prevent resistant microbes from emerging or surviving selective pressure by an antimicrobial agent.

Modeling the effect of antimicrobial agents on heterogeneous microbial populations: reducing complexity by categorization

When an agent is used on a microbial population that contains microbes resistant to that agent, part of that population will survive and grow, while the rest of the population will be eliminated. There are multiple mechanisms, which impart antimicrobial resistance, involving single or multiple functions that interact in non-trivial ways. Moreover, the interactions among a microbial population, infected host, and antimicrobial agent impart additional complexity, in the sense that they involve an overwhelmingly large number of factors which can hardly be monitored and whose interplay is poorly understood.

Nevertheless, the apparent behavior of a single kind of microbes in the presence of an antimicrobial agent is either survival or elimination. Therefore, the dichotomic categorization of microbes as resistant and susceptible can be considered as a first step towards reducing modeling complexity for a bacterial population. Quantitative models have been constructed that separate a population into resistant and susceptible subpopulations, and attempt to capture the quantitative effect of an agent on the entire population in terms of differential equations corresponding to standard population balances (Giraldo, Vivas, Vila, & Badia, 2002; Lipsitch & Levin, 1997).

Limitations of The Resistant/Susceptible Dichotomy

Background

Consider a homogeneous microbial population of N_0 microbes in an environment of antimicrobial agent concentration C. Under the assumption that all microbes have the same susceptibility to the antimicrobial agent at concentration C, one can write the standard population balance equation (Wagner, 1968 and Jusko, 1971).

$$\frac{dN}{dt} = \underbrace{K_g N(t)}_{\text{physiological growth rate}} - \underbrace{r(C)N(t)}_{\text{kill rate due to agent}}.$$

$$N(0) = N_0 \qquad (16)$$

where N(t) is the total number of microbes at time t; Kg is the physiological growth rate per unit of the microbes (net effect of natural microbial growth and death); r(C) is the antimicrobial agent-induced kill rate per unit of microbes, which is a non-decreasing function of the antimicrobial agent concentration C; and the approximation $K_g N(t) \approx K_g [(1 - (N(t)/N_{max}(t))] N(t)$ is used for the physiological growth rate when the microbial population is not growing towards its saturation value N_{max}. The form of r(C) depends on the kind of microbes and antimicrobial agent involved. A fairly widely used expression (Giraldo et al., 2002) is

$$r(C) = \frac{K_k C^H}{C^H + C_{50}^H} \qquad (17)$$

where C is the antimicrobial agent concentration; Kk is the maximal kill rate achieved as C → ∞; C_{50} is a constant equal to the antimicrobial agent concentration at which 50% of the maximal kill rate is achieved; and H is the so-called Hill factor, corresponding to how inflected r is as a function of C. The analysis in the sequel is general and is not dependent on the particular functional form of r(C).

For a time-invariant concentration C the standard solution of Eq. (16) is $\ln[N(t)/N0] = \underbrace{(K_g - r(C))t}_{\alpha}$, indicating an increasing or decreasing straight line in a semi-logarithmic plot of N(t) versus t, depending on the sign of Kg – r(C). When Kg – r(C) = 0 the microbial population will neither grow nor decline but will remain constant.

The antimicrobial agent concentration that achieves this effect is the routinely and widely used minimal inhibitory concentration (MIC) (Craig, 1998; Mueller, de la Peña, & Derendorf, 2004; National Committee for Clinical Laboratory Standards, 1997). Therefore, the value of r(C) in comparison with Kgrepresents the resistance of microbes to a specific antimicrobial agent at concentration C.

While Eq. (16) can capture the time-dependent profiles of monotonically growing or declining populationsN(t), it cannot capture situations where a microbial population experiences initial decline followed by later growth. To capture this phenomenon it is customary to consider two different subpopulations of different resistance each, and apply equations such as Eq. (16) to each subpopulation. This approach is conceptually appealing and computationally simple, but it is only a rough approximation of the real system that may fail to predict phenomena such as regrowth of resistant microbes, as we show below. As we will explain in the sequel, population resistance is distributed over a multitude (virtual continuum) of values rather than over two values only.

Motivating Example

Consider the data shown in Fig. 20.

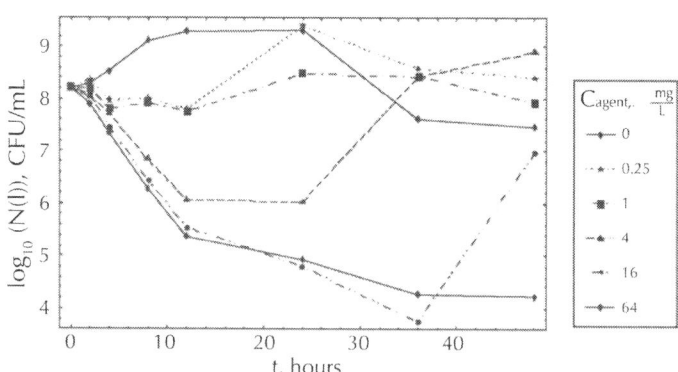

Figure 20: Effect of meropenem (a carbapenem antibiotic) on Pseudomonas aeruginosa ATCC 27853. (Tam, Schilling, Melnick & Coyle, 2005).

It is evident from the regrowth patterns (i.e. lack of straight lines) observed in Fig. 20 that resistance to the antimicrobial agent is not homogeneous over the entire population. One can consider two subpopulations, one of resistant and one of susceptible microbes, and write standard equations similar to Eq. (16) for each subpopulation, to finally get

$$\ln \frac{N(t)}{N_0} = \alpha t + \ln[(e^{\beta t} - 1)\gamma + 1]$$

(18)

where $\alpha = \hat{K}_g - r_{susc}(C)$, $\beta = \hat{r}_{susc}(C) - r_{res}(C)$, $\gamma = \hat{(N_{res,0}/N_0)} - 1$. Applying non-linear least squares (Mathematica® function Nminimize, Differential Evolution algorithm) with Eq. (18) and the first 24 h of the data in Fig. 20 for agent concentrations C = 0.25 mg/L through C = 64 mg/L we got parameter estimates for α, β, and γ for each C.

Plugging the above parameter estimates into Eq. (18), we produced the curves of Fig. 21, showing the total microbial population as a function of time for time 0 to 48 h. While these curves fit the data well, they fail to predict the regrowth experienced at antimicrobial agent concentrations C = 4,16,64 mg/L mg/L over a period of 48 h.

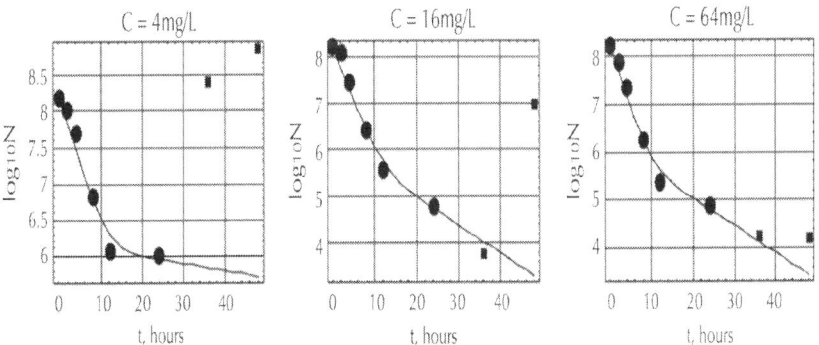

Figure 21: Comparison between predictions via Eq. (18) and measurements (squares) at 36 and 48 h. The fraction of the resistant bacteria subpopulation at t = 0 was estimated to be $N_{res,0}/N_0 = 0.0095$.

Therefore, the simple division of the entire microbial population into two subpopulations, a resistant one and a susceptible one, clearly fails to model this system adequately, in that it fails to predict regrowth, which is a key reason why subpopulations are considered in the first place.

To overcome these difficulties we summarize below a novel approach, founded on the assumption that the resistance of a microbial population is distributed over a virtual continuum rather than over two distinct values. Details can be found in (Nikolaou & Tam, 2005).

A New Approach to Dynamics of Heterogeneous Microbial Populations

Consider now a heterogeneous microbial population in which resistance to an antimicrobial agent varies among microbes, with more resistant microbes corresponding to lower kill rates r(C).

Theorem 1.

Dynamics of kill rate in heterogeneous microbial population

Assume that a heterogeneous population of N_0 bacteria is subjected to constant antimicrobial agent concentration C. The population has a number of subpopulations, each of which satisfies Eq. (16). Then

a. The entire bacterial population evolves as

$$\frac{dN}{dt} = (K_g - \mu_r(t))N(t) \tag{19}$$

where μr(t) is the average kill rate over the entire population at time t.

b. The average kill rate μr(t) evolves as

$$\frac{d\mu_r}{dt} = -\sigma_r(t)^2 \tag{20}$$

where $r(t)^2$ is the variance of the kill rate distribution over the population at time t.

c. The variance $r(t)^2$ of the kill rate distribution evolves according to the equations

$$\frac{d\sigma^2}{dt} = -\kappa_{r,3}(t) \qquad (21)$$

$$\frac{d\kappa_{r,n}}{dt} = -\kappa_{r,n+1}(t), \quad n \geq 3 \qquad (22)$$

where the initial conditions of the variable κr,n are the cumulants (Papoulis, 1984) of the initial kill rate distribution. (Proof in Nikolaou & Tam, 2005).

For initial distributions of kill rate that are close to normal, the variance σr(t)² of the kill rate remains approximately constant for a certain period of time, i.e.

$$\sigma_r(t)^2 \approx \sigma_r(0)^2 \equiv \sigma_r^2 \qquad (23)$$

Theorem 2.

Explicit expressions for heterogeneous microbial population

The approximation of Eq. (23), combined with Eqs. (19) and (22), yields the following expressions for time t less than

$$t_{max} = \frac{\mu_r(t)}{\sigma_r^2} \qquad (24)$$

a. The average kill rate declines as

$$\mu_r(t) \approx \mu_r(0) - \sigma_r^2 t \qquad (25)$$

b. The entire population evolves as

$$\ln\left[\frac{N(t)}{N_0}\right] \approx \underbrace{(K_g - \mu_r(0))t}_{a} + \underbrace{\tfrac{1}{2}\sigma_r^2 t^2}_{b} \qquad (26)$$

(Proof in Nikolaou & Tam, 2005.)

Eqs. (25) and (26) are remarkably powerful in that they describe the time profile of an entire microbial population with distributed resistance in terms of a model with only two parameters (in addition to Kg which can be easily obtained from growth experiments alone): the initial average kill rate $\mu_r(0)$ and its variance σ^2_r. It is clear that the parameters a and b can be trivially estimated from experimental data using linear regression. Eq. (24) in Theorem 2 is not restrictive, because our main interest is in detecting whether regrowth will occur on the basis of scant experimental data. This question can be answered by the following.

Theorem 3.

Elimination or regrowth of an entire microbial population

The entire microbial population will be eliminated approximately at time

$$T_{\text{elimination}} \approx \frac{\mu_r(0) - K_g - \sqrt{(\mu_r(0) - K_g)^2 - 2\sigma_r^2 \ln N_0}}{\sigma_r^2} > 0 \qquad (27)$$

if, approximately,

$$\mu_r(0) - K_g - \sigma_r\sqrt{2\ln N_0} > 0 \qquad (28)$$

Otherwise, the population will start regrowing approximately at time

$$T_{\text{regrowth}} \approx \frac{\mu_r(0) - K_g}{\sigma_r^2} \qquad (29)$$

(Proof in Nikolaou & Tam, 2005.)

The preceding results suggest the following simple methodology, via Eq. (28), for the determination of an antimicrobial agent concentration that will eliminate all microbes, including the most resistant ones, and will avoid regrowth. On the basis of Eq. (28), an antimicrobial agent concentration C is sought which will ensure that

$$f(C) \hat{=} \mu_r(0) - K_g - \sigma_r \sqrt{2 \ln N_0} \qquad (30)$$

where the quantities $\mu_r(0)$ and σ_r in the above Eq. (30) depend on the antimicrobial agent concentrationC, as discussed in Theorems 1 or 2. To determine the value of C that satisfies Eq. (30), one can conduct experiments at different concentrations, curve-fit $\mu_r(0)$ and σ_r as functions of C, then simply plot f(C) versus C and check when f(C) = 0.

Case Study

The ideas discussed in the preceding Section 3.4 were applied to the same experimental data of Section3.3, shown in Fig. 20. Our objective was to detect whether regrowth would occur by analyzing data collected over a 24-h period (dots in Fig. 22). To do that, we used linear regression to estimate a and b in Eq. (26) from 24-h data. Then we used the estimated values of a and b and applied Eq. (26) beyond 24 h to predict the total population (squares in Fig. 22) up to the time indicated by Eq. (24). Note that Eq. (24)necessitates an estimate of $\mu r(0)$, which was obtained from knowledge of a and Kg, where the latter was estimated from the pure growth experiment shown in Fig. 20 corresponding to C = 0 mg/L.

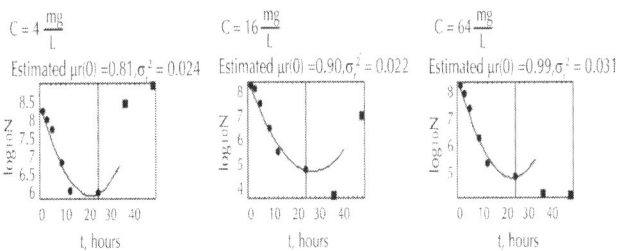

Figure 22: Comparison between predictions via Eq. (26) and measurements (squares) at 36 and 48 h for data in Fig. 20.

In contrast to the two-subpopulation model, which failed to predict regrowth (Fig. 21), the proposed model predicted regrowth very well for C = 16 mg/L (Fig. 22d). For C = 64 mg/L the proposed model predicted that regrowth would occur, albeit the predicted time was not accurate. Since data beyond 48 h were not available, it was not possible to rigorously assess the validity of that prediction at this time.

Fig. 23a shows the initial average kill rate estimate μr(0) as a function of the antimicrobial agent concentration C according to the data shown in Fig. 22(dots) and curve fitting by an expression similar to Eq. (17). Similarly, Fig. 23b shows the variance of the kill rate as a function of the antimicrobial agent concentration C.

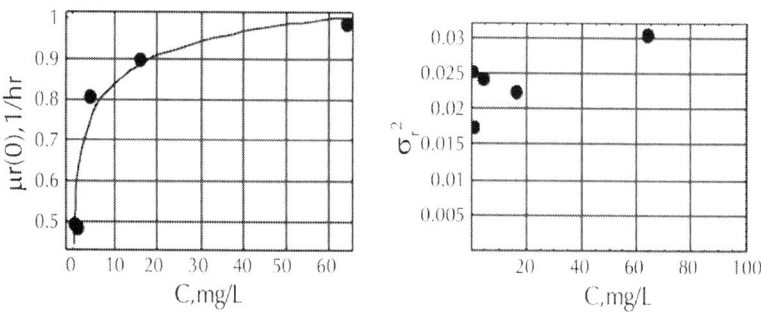

Figure 23: Estimates of initial kill rate average μr(0) and variance σ_r^2 based on the first 24 h of data shown in Fig. 22 for the case study of Section 3.5. Continuous line: Curve fitting by an expression similar to Eq. (17) with $\overline{C}_{50} = 1.4$, H = 0.36, Kk = 1.26.

Fig. 24 shows the value of $f(c) \triangleq \mu_r(0) - K_g - \sigma_r\sqrt{2\ln N_0}$, required by Eq. (28) to be positive for elimination of all microbes. Clearly f(C) is not positive for the values of the agent concentration C considered, indicating that larger values must be used for complete elimination of all microbes. What value of C would guarantee this could not be accurately predicted from the available data, due to various approximation errors in the equations themselves as well as in the data. This is an issue that will be further explored.

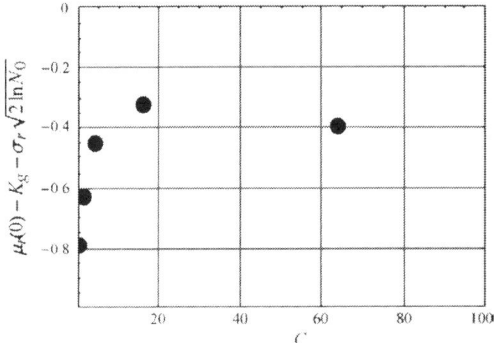

Figure 24: Test of satisfaction of Eq. (28) to determine at what C regrowth will be avoided for the case study of Section 3.5.

Summary: The Complex Interaction between Microbes and Antimicrobial Agents

We have discussed a method for detecting whether regrowth will occur in a microbial population under antimicrobial agent pressure in standard in vitro experiments. The method relies on differential equations that we developed for the total number of microbes, for the average kill rate, and for the kill rate variance (Eqs. (19) through (21)). Under the simplifying assumption that the kill rate variance remains temporarily constant over time (Eq.(23)) a simple analytical expression can be obtained for the total number of microbes as a function of time, involving only two parameters (beyond the physiological growth rate Kg): the average and variance of the kill rate of the initial population. These parameters can be very easily estimated by linear regression. The proposed method was successful in predicting regrowth for a microbial population for which data beyond 24 h were available for validation of predictions. By contrast, a two-subpopulation model with parameters fitted using the same data failed to predict regrowth.

We want to emphasize that even though the proposed approach is general, what we have presented here is only its basic framework. There is a number of issues that can be further explored to improve and expand the method, such as the following:

-While quite realistic for many systems, the approximation provided by Eq. (23) is clearly valid only for a certain period of time. How can the exact counterpart of Eq. (23), namely Eqs. (21) and (22) be used for even more accurate predictions?

-All results presented in this work assume the standard in vitro setting of a time-invariant agent concentration C. The more interesting case of C varying in a way that mimics in vivo patterns (corresponding to living organism pharmacokinetics) would provide useful insight into the design of agent dosing regimens that would prevent regrowth due to resistant microbes. It would also help the development of new agents, by making predictions that would drastically cut the number of trials required for new agent development.

-It would be an understatement to say that real systems are far more complex than the preceding discussion implies. For example, spontaneous mutations can create resistant microbes in the course of action an antimicrobial agent. Detecting and preventing such phenomena would clearly require an ambitious extension of the framework proposed here, and is crucial for the design of effective antimicrobial agents and treatments.

DRILLING OF HYDROCARBON-PRODUCTION WELLS

The Importance of Maintaining Optimal Weight on Bit (WOB)

Maintaining optimal weight on the bit of a hydrocarbon-production well drilling system is of great economic significance. As shown in Fig. 25, a typical hydrocarbon-well drilling system relies on the rotation of a drill bit (Fig. 26) – attached to the drill string at its lower end – to penetrate a rock formation, in order to reach a hydrocarbon reservoir (Economides, Watters, & Dunn-Norman,

1988). Drilling proceeds at the rate of a few meters to a few tens of meters per hour and is quite expensive. In fact, it is the single most expensive activity in the development of an oil field. Therefore, achievement of high (ideally maximum) rate of penetration (ROP) is of great importance when drilling hydrocarbon wells (Guo, 1988 and Kennedy, 1983). To achieve good ROP into a given layer of a rock formation, it is important that a drill bit is rotated at the speed for which it was designed, and the weight placed on the drill bit (weight on bit, WOB) is close to (preferably at) an optimal value (Fig. 27). Consequently, one can attempt to enhance (ideally optimize) ROP by ensuring optimal weight on bit (Boyadjieff, 1986, Cunningham, 1978 and Gatlin, 1957; Jorden & Shirley, 1966). This task entails

(a) Knowing what the optimal WOB is (a function mainly of the rock formation and bit design) – a task nowadays purportedly achievable by commercial products (e.g., Noble Corporation, 2005) –, and

(b) Controlling WOB at its (estimated) optimal setpoint value by adjusting the feed rate of the steel cable spooled from the cylindrical drum in the drawworks (Fig. 25).

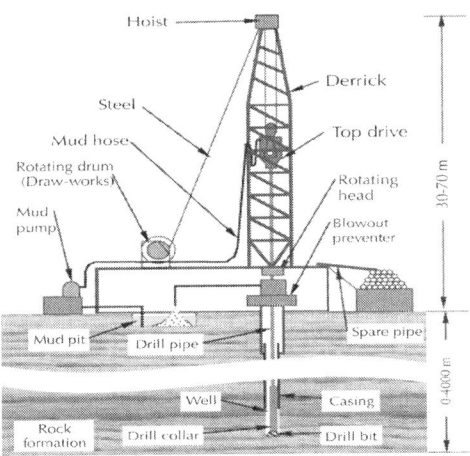

Figure 25: Elements of a drilling rig. A tall derrick is used to suspend a top-drive motor to which a long drill string (in red) is attached. The drill

string is made by adding sections of steel drill pipe (~9 m long, ~10 cm OD) as the hole is being drilled into the rock formation. At the lower end of the drill string a thicker, rigid, heavy drill collar is attached. The drill collar concentrates weight and rigidity at the lower end of the drill string as the drill string gets longer. The drill string transfers rotation from the top-drive (Boyadjieff, 1986) (or from arotary table) to the drill bit at the bottom. Additional bit rotation is generated by a turbine-like bottom-hole motor operated by drilling mud, a complex mixture of water and special clays. Mud is pumped down the borehole through the drill string and motor and then up through the annulus. Drilling mud functions also include lubrication and cooling of the bit; control of pressure; removal of cuttings; and transmission of bottomhole measurements as pulse signals to the surface. The blowout preventer is a series of safety valves that help prevent any oil, gas, or water from shooting to the surface.

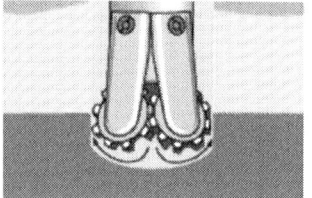 

Figure 26: Roller-cone (School Science, 2005) and polycrystalline diamond compact (PDC) (Baker Hughes, 2005) drill bits used extensively in most drilling rigs.

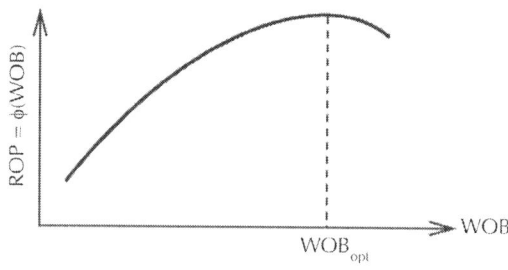

Figure 27: The ROP of a drill bit into a rock formation is a monotonically increasing function of WOB (Cunningham, 1978 and Gatlin, 1957) up to an optimal point, beyond which increasing WOB decreases ROP

(Brett, Warren, & Wait, 1990; Eronini, Somerton, & Auslander, 1982) to the point of eventually inducing catastrophic failures. In addition, the value of WOB at which maximum ROP is attained is a strong function of the inhomogeneities of the rock formation being drilled, and one that cannot be accurately known ahead of time. The decline of ROP for large WOB is due to a number of factors, such as drilling inefficiencies, poor removal of cuttings, etc.

WOB fluctuations around the WOB optimal value (even below ±1%) are costly, not only because they result in lower ROP (Fig. 27), but also because excessive WOB causes premature wear or catastrophic failure of the drill bit, thus forcing its early replacement - an expensive proposition (Spanos, Chevallier, Politis, & Payne, 2003). Owing to technological limitations, field practice has been to manually or automatically control WOB somewhat below its optimal value. Recent technological advances, such as bottomhole sensors, data acquisition systems, mechanical hardware, and abundantly inexpensive computing power have spurred attempts to control WOB at its optimum value and thereby realize significant economic benefits (Boyadjieff, Murray, Orr, Porche, & Thompson, 2003). In addition to successes, such attempts have also met serious challenges. Maintaining optimal weight on the bit of a hydrocarbon-well drilling system is fundamentally challenging. As we explain below, there is a fundamental reason – among many practical ones (Boyadjieff, 1986) – why controlling WOB at its optimal value is challenging, summarized as follows: If the drill bit has to operate at a steady state corresponding to optimal WOB, then that steady state is at the interface of stable and unstable steady states, corresponding to WOB lower and higher than the optimum WOB, respectively (Fig. 27). Therefore, a control system that maintains WOB at its optimum value must stabilize a non-linear system that exhibits both locally stable and locally unstable behavior.

The problem is exacerbated by

-uncertainty in both the drilling system and the properties of the rock formation being drilled, which render the effect of WOB on ROP highly uncertain;

-inhomogeneity of the rock formation being drilled, which makes the effect of WOB on ROP time-varying as drilling proceeds;

-constraints on process variables; and

-real-time measurement delays.

Therefore fundamental analysis and synthesis of proposed solutions must be undertaken before the potential of commercial development and implementation of optimal WOB control is fully realized. Such work can proceed along the following lines:

-Use a combination of first principles and published field observations to conduct a fundamental analysis of the dynamic behavior of a drilling system that operates close to its optimal WOB;

-Develop and analyze the properties of novel control strategies that can be used to control WOB close to its optimal value, in the presence of

- Stability/instability non-linearities,
- Uncertainty,
- Time-variability,
- Constraints;

-Perform computer simulation experiments to conduct extensive testing of the control strategies proposed above at the proof-of-concept level; and

-Suggest how to develop software and hardware tools that can be practically applied in the field.

We present some preliminary analysis below.

Preliminary Results and Research Framework

A Simplified Drill String Model Reveals Rich Dynamic Behavior Close to Optimal WOB

For the sake of exposing fundamental issues with control of WOB at its optimal value, let us consider a very simplified model of the drill string under tension, shown in Fig. 28. Changes in drill string tension, and all rotation effects are temporarily ignored.

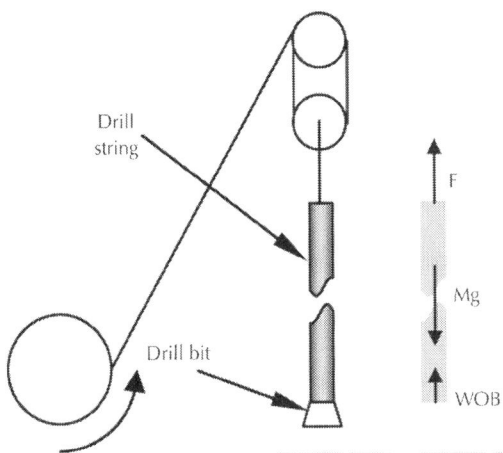

Figure 28: A simplified model of drill string indicating force balance. Compare with Fig. 25.

A standard balance of forces on the drill string yields

$$M\frac{d(\text{ROP})}{dt} = Mg - F - (\text{WOB}) \tag{31}$$

where M is the drill string mass; ROP is the rate of penetration; g is the acceleration of gravity; F is the sum of all other forces on the drill string, e.g. hook load and friction; and WOB is the weight on bit. The weight on bit, WOB, and rate of penetration, ROP, are related by a non-linear relationship of the form

$$\left(a_n\frac{d^n}{dt^n} + \cdots + a_1\frac{d}{dt} + a_0\right) ROP = \phi(WOB) \tag{32}$$

where $a_i \ll$, and the function $\phi(\text{WOB})$ is non-linear and exhibits a maximum at an optimal value of weight on bit, WOB_{opt}, as shown in Fig. 27. It can then be shown (Awasthi & Nikolaou, submitted for publication) that Eqs. (31) and (32) imply

$$\left(a_n \frac{d^n}{dt^n} + \cdots + [a_1 + Ma(\text{WOB}_{opt} - \text{WOB})]\frac{d}{dt} + a_0\right) \text{WOB}$$

$$\approx \left(a_n \frac{d^n}{dt^n} + \cdots + a_1 \frac{d}{dt} + a_0\right)(Mg - F) \tag{33}$$

in the neighborhood of WOB_{opt}, where $a = -(d^2\phi/d(\text{WOB})^2)|_{\text{WOB}_{opt}}$.

Because the coefficient $[a1 + Ma(\text{WOB}_{opt} - \text{WOB}_{ss})]$ in Eq. (33) changes from positive to negative when WOB_{ss} increases to cross the critical value

$$\text{WOB}_{cr} = \text{WOB}_{opt} + \frac{a_1}{Ma} \approx \text{WOB}_{opt}, \tag{34}$$

the behavior of the non-linear system of Eq. (33) is locally stable around steady-state values of WOB approximately below WOB_{opt} but is unstable around steady-state values of WOB approximately above WOB_{opt} (Fig. 29) according to the Routh–Hurwitz stability criterion. Therefore, if feedback is to maintain WOB around its optimal value WOB_{opt}, then it must stabilize a non-linear system that exhibits both locally stable and locally unstable behavior.

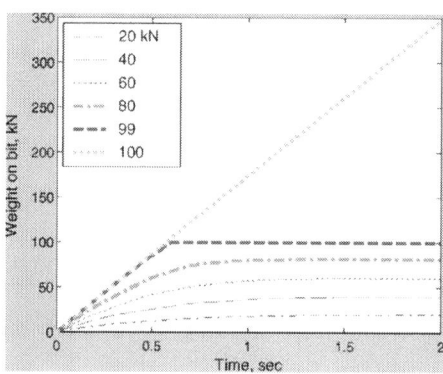

Figure 29: Response of WOB to the changes in the force Mg-F (Eq. (33)) shown in the legend. Note the non-linearity of the responses and the tran-

sition to instability when the sought steady state of WOB is at the optimal value $WOB_{opt} = 100$ kN (Fig. 27).

Note that the above conclusion is independent of

(i) the detailed form of the function ɸ(WOB) in Fig. 27 – as long as ɸ exhibits a maximum – and
(ii) the level of detail used to model the drill string and particular geometry of the borehole.

Indeed, the cause of the above stability/instability behavior is that a lower part of the drill string (e.g., almost rigid drill collar and drill bit) experiences ROP through a function ɸ(WOB) that attains a maximum.

Simple Strategies Cannot Easily Control WOB Close To Its Optimal Setpoint

While it would be desirable to keep the complexity of a WOB control algorithm as low as possible, the dynamics of Eq. (33) make the problem inherently challenging. At the very least, a controller would be required to simultaneously stabilize a pair of a stable and an unstable system, corresponding to positive and negative values of the coefficient $[a_1 + Ma(WOB_{opt} - WOB_{ss})]$ in the local linearization of Eq. (33). This is an instance of the problem of simultaneous stabilization of linear plants, which has been extensively studied in literature (Blondel, 1994; Blondel and Tsitsiklis, 2000). It is now well known that the rigorous solution of the simultaneous stabilization problem is inherently "hard", i.e. variants of the problem are NP-hard, NP-complete, or undecidable. (In fact, as an alternative to a rigorous solution, randomized algorithms, which are required to work "most of the time", rather than "all the time", have been proposed,Vidyasagar, 2001). For example, if $a_1 = 0$, it is trivial to show that no P controller can stabilize Eq. (33). In addition, tuning a PI controller is all but trivial, owing to the uncertainty in the optimal value of WOB. Fig. 30shows that closed-loop response may change from excellent to unstable for WOB setpoints going from below the optimum WOB to just 2% above it.

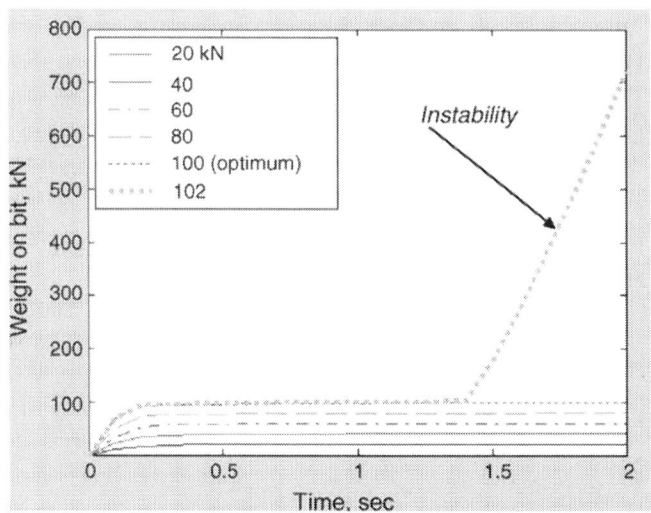

Figure 30: Closed-loop response of WOB to step changes of increasing magnitudes in its setpoint when the PI controller $C(s) = 4(1 + (1/0.5\ s))$ is used (Nikolaou, 2003). While all responses for setpoint changes from 20 to 100 kN are stable, the response to a setpoint change of magnitude 102 kN is unstable. Incidentally, note that the closed loop is much more linear than the open loop (Fig. 29) for setpoint changes that do not result in instability (100 kN and below).

There are additional elements that complicate the problem, such as the following:

-Model uncertainty: The function $\phi(WOB)$ in Fig. 27 and the coefficients a_1 in Eq. (32) are uncertain. In fact, ϕ fluctuates with time according to the consistency of the rock formation. Therefore, neither the actual setpoint of WOB_{opt} nor the dynamic model of Eq. (33) is known with certainty.

-Constraints on manipulated inputs: The force F, which can be manipulated by the drawworks is bounded, according to the maximum braking power available at the drawworks.

-Constraints on controlled outputs: To maintain ROP at a fairly high value and avoid premature bit wear, WOB must remain within bounds, e.g. ±2000 N from WOB_{opt} (Fig. 27). "Anti-birdnesting" constraints on acceleration are also present (Boyadjieff et al., 2003).

-Delays causing lack of synchronization of WOB and ROP measurements.

Potential Control Strategies

To address the preceding control challenges, one can develop an efficient non-linear model predictive control (NMPC) approach. While constrained model predictive control (MPC) was first applied in the 1970s to relatively slow chemical processes and is now the de facto standard of advanced multivariable control in process industries (Nikolaou, 2001 and Qin and Badgwell, 1997; Qin & Badgwell, 2000), it is finding applicability in many other application areas (Morari, 2002). MPC with a linear model is a fairly mature topic (Morari & Lee, 1999; Mayne, Rawlings, Rao, & Scokaert, 2000; Nikolaou, 2001 and Rawlings, 2000). In addition, significant progress has been made on MPC with non-linear process model (Allgöwer F & Zheng, 2000; Berber & Kravaris, 1998). Therefore, NMPC is a promising candidate control strategy for optimal WOB control.

Setting up the control problem as a non-linear on-line optimization subject to constraints is straightforward. Solution of the on-line optimization is not trivial, because of non-convexities due to the term $(WOB_{opt} - WOB)(dWOB/dt)$ in Eq. (33). Repeated use of local linearization ideas may prove helpful. State estimation using a moving horizon approach (Rao, Rawlings, & Lee, 2001) may also be fruitful.

The efficiency of on-line control computations is important for practical implementation of NMPC. Given that sampling rates are above one per second, one can to investigate how the on-line optimization can be

(a) performed efficiently, or

(b) replaced by a closed-form control law.

The first direction (a) suggests development based on combining the model structure of the controlled system (e.g., Eq. (33)) with numerical on-line optimization ideas, such as

-efficient parametrization of decision variables (system inputs);

-exploitation of NMPC-specific structure, e.g. automatic discarding of redundant constraints at later parts of the optimization horizon (Biegler, 1998; Chen and Allgöwer, 1998; Henson, 1998; Findeisen & Allgöwer, 2003);

-use of information from the optimal solution of the previous time point, to "warm start" the optimization at the following time point;

-exploitation of the optimization problem structure (several linear terms) (Biegler, 1998; Biegler & Sentoni, 2000).

The second direction (b) is associated with the applicability of multiparametric programming. This tool has been successfully used to replace linear MPC by closed-form control laws, which, at each moment, use the state of the controlled system to consult a look-up table which selects a time-invariant controller from a given set (Bemporad, Morari, Dua, & Pistikopoulos, 2002). To develop a similar approach for the non-linear problem at hand, one can apply the Karush–Kuhn–Tucker optimality conditions to the on-line optimization problem and exploit its structure (few parameters change from one time point to the next, most constraints are linear) in order to identify state-space polytopes corresponding to the same set of active or inactive constraints. In contrast to linear MPC, these polytopes will include non-linear boundaries in NMPC.

Controller Design in ihe Presence of Uncertainty

Even though the function $\phi(WOB)$ (Fig. 27) can be estimated on-line, uncertainties will exist. In fact, critical information about the function $\phi(WOB)$ is captured by WOB_{opt} as well as $\phi(WOB_{opt})$ (see Eq. (33)). The effect of uncertainty in WOB_{opt} on control performance is expected to be stronger than the effect of the value of $\phi(WOB_{opt})$. Options to examine that mitigate the effect of uncertainty (increase robustness) for the on-line-optimization-based form of NMPC can be summarized as:

(a) Tuning the on-line optimization objective, including weights and horizon lengths;
(b) Augmenting the NMPC structure by adding (essentially Lyapunov-type) constraints that ensure robustness; and
(c) Including appropriate terminal constraints and penalties.

Assessment and Reduction of Uncertainty

Even though there exist commercially available solutions for on-line estimation of $\phi(WOB)$, the interplay of such solutions with optimal WOB control merits thorough investigation, because of the direct coupling between identification and feedback control. Indeed, it is clear that if a system is inadvertently controlled around a set value of WOB, then knowledge of $\phi(WOB)$ will obviously remain limited. On the other hand, attempting to identify $\phi(WOB)$ might drive the system to WOB values far from the optimum, owing to (a) poor initial guess of $\phi(WOB)$ and/or (b) poor control because of poor knowledge of $\phi(WOB)$. Therefore, coordination of the identification and control activity warrants thorough investigation.

Improved Modeling

As already stated in Section 4.2.1, the model of Fig. 25 is an extreme simplification of the actual system, which nevertheless reveals the fundamental stability/instability issues that have to be dealt with when controlling WOB close to its optimum. It is clear that this model can be improved. For example, a straightforward improvement that takes into account the elasticity of the drill string is shown in Fig. 31. As explained in Section 4.2.1, this will not invalidate the fundamental stability/instability issues raised above.

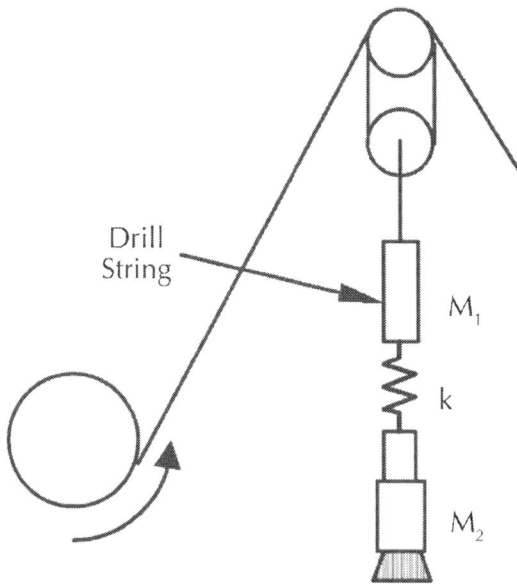

Figure 31: A simple first step towards the refinement of the drill string model. Compare with Fig. 25 and Fig. 28.

In addition to the preceding refinement, other refinements can be considered. Along these lines, Spanos and co-workers (2003) have provided a thorough review of several modeling issues, such as ROP modeling, torque and drag modeling, bit-bounce phenomena, stick-slip phenomena, etc. Including and studying all of these phenomena in a model would be overly complex. Therefore, phenomena can be considered selectively, to the extent of their impact on WOB control.

Extension to Directional Drilling

Directional drilling (Fig. 32) has grown dramatically in the last several years. While the principles outlined in the previous sections (e.g. the fundamentals outlined in Section 4.2.1) are directly applicable to directional drilling, there is much higher level of complexity associated with it. In particular, the following issues can be addressed:

-WOB in directional branches cannot be measured directly (because strictly speaking it is not "weight" but the force acting on the bit), and is usually inferred from the pressure drop across the mud motor at the end of the drill string. In addition to inaccuracies, this creates measurement delays of several seconds due to limited bandwidth in the transmission line for encoding pulses traveling through the mud to the surface.

-Because the drill string is not straight, well-bore tortuosity and mud properties (in addition to rotating speed and lithology) now have a strong effect on frictional forces, cf. Eq. (31).

-Partial differential equations (usually discretized as a series of ordinary differential equations) describe the dynamics of the drill string. The basic Eq. (31) will now refer to the last element of the drill string rather than to its total length. The entire drill string is partitioned to a number of torque and drag elements, and balances are written for each one of them.

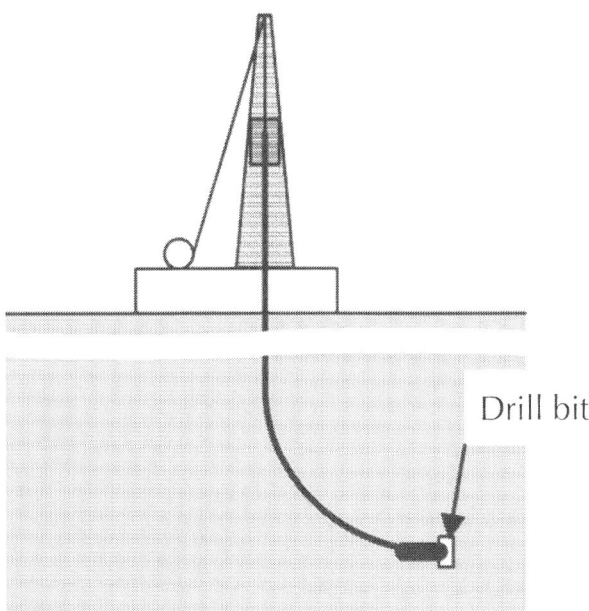

Figure 32: Schematic of a directional drilling rig.

Summary: Complexity in Hydrocarbon-Production Well Drilling

The interaction between drilling equipment and rock formation creates dynamic behavior that is far more complex that that of either of two constituents in isolation. This creates interesting challenges for controlling WOB at its optimum. Certainly, WOB control is only one element of automated drilling operations (Maidla, 2002; Saputelli, Economides, Nikolaou, Demarchos, & Sagias, 2003). Recent experience indicates that in order to optimize the drilling operation the entire drilling system, not just the mechanics or software, needs to be designed from a control system point of view. For example, adjusting the manipulated input (part of the force F in Eq. (31)) used to control WOB may eventually require a linear braking effect from the brake in the drawworks (Fig. 25), a task made difficult by the very design of widely used self-energizing band brakes. Wide automation of drilling operations is a future vision, which current trends indicate to be realizable (Saputelli et al., 2000; Saputelli, Malki, Canelon, & Nikolaou, 2002).

DISCUSSION

We have presented studies on complex behavior in three areas: semiconductor manufacturing, activity of antimicrobial agents, and drilling of hydrocarbon wells. While the physical systems involved in these studies are completely different, there are common themes in the cause and treatment of complexity.

-System behavior is not merely the sum of individual behaviors.

- In the etching of silicon wafers, the reactor chamber, wafer, and plasma interact in ways that make it difficult to configure the reactor and adjust operating conditions (recipe) for uniform etching of wafers.
- When assessing the effect of antimicrobial agents on microbes, interactions among agent, host, and microbes make it difficult to extrapolate in vitro data to realistic in vivo conditions.

- The interaction between drilling equipment and rock formation creates non-linear behavior that cannot be readily observed in either factor.

-Dealing with complexity may rely on successful model simplification (reduction) and focusing on key aspects of the behavior of a system.

- The effect of different reactor configurations and operating conditions (recipes) on etching uniformity can be captured by focusing on a couple of characteristic patterns of wafer-wide etching that can be experimentally determined.
- The effect of antimicrobial agents on heterogeneous microbial populations can be captured by focusing on a few crucial parameters (e.g., average and standard deviation of kill rate) of a microbial population treated by an agent.
- While high-fidelity models are available for simulation of drill-string dynamics via finite-element analysis, the effect of weight-on-bit on the rate of penetration into a rock formation can be captured by a single differential equation that is crucial for the behavior of the system.

As a recent trend in science and engineering is to study systems rather than individual components, "complexity" issues (in the sense of resultant behavior emanating from the interaction of individual components in a system) are bound to emerge in many places and to offer stimulus for creative work.

In closing, we want to iterate our claim in the Introduction that an all-encompassing definition of complexity may be elusive and not entirely necessary. Perhaps a Supreme Justice's much quoted criterion may be appropriate: "You can't define it, but you know it when you see it."

ACKNOWLEDGEMENTS

The authors gratefully acknowledge partial support for this work from the following organigations: Lam Research Corporation, Fremont CA for providing partial funding and experimental facilities;

National Science Foundation through grants SGER-0118516 and CTS-36525000; University of Houston GEAR program; American Chemical Society—Petroleum Research Fund and Noble Drilling for many fruitful discussions.

REFERENCES

1. Allgower F., & Zheng, A. (Eds.). (2000). ¨Nonlinear predictive control. Birkhauser.
2. Ash, C. (1996). Antibiotic resistance: the new apocalypse? Trends in Microbiology, 4, 371–372.
3. Astr¨om, K. J. (2001). Introduction. In K. J. ¨Astr¨om, P. Albertos, M. Blanke, ¨A. Isidori, W. Schaufelberger, & R. Sanz (Eds.), Control of complex systems (pp. 1–20). Springer.
4. Awasthi, A., & Nikolaou, M. (submitted for publication). Controlling weight on bit at its optimum in hydrocarbon well drilling operations. AIChE Journal.
5. Baker Hughes. http://www.bakerhughes.com/bakerhughes/casefile/measurecompare genesis.htm.
6. Bemporad, A., Morari, M., Dua, V., & Pistikopoulos, E. (2002). The explicit linear quadratic regulator for constrained systems. Automatica, 38(1), 3–20.
7. Berber, R., & Kravaris, C. (Eds.). (1998). Nonlinear model based process control. Kluwer.
8. Bharati, M. H., Liu, J. J., & MacGregor, J. F. (2004). Image texture analysis: methods and comparisons. Chemometrics and Intelligent Laboratory Systems, 72(1), 57–71.
9. Bharati, M. H., & MacGregor, J. F. (1998). Multivariate image analysis for real-time process monitoring and control. Industrial & Engineering Chemistry Research, 37(12), 4715–4724.
10. Bharati, M. H., MacGregor, J. F., & Tropper, W. (2003). Softwood lumber grading through on-line multivariate image

analysis techniques. Industrial & Engineering Chemistry Research, 42(21), 5345–5353.
11. Biegler, L. (1998). Advances in nonlinear programming concepts for process control. Journal of Process Control, 8(5–6), 301–311.
12. Biegler, L., & Sentoni, G. (2000). Efficient formulation and solution of nonlinear model predictive control problem. Latin American Applied Research, 30(4), 315–324.
13. Blondel, V. (1994). Simultaneous stabilization of linear systems. Springer.
14. Blondel, V., & Tsitsiklis, J. (2000). A survey of computational complexity results in systems and control. Automatica, 36(9), 1249–1274.
15. Boyadjieff, G. (1986). An overview of top-drive drilling system applications and experiences. SPE Drilling Engineering, 435–442.
16. Boyadjieff, G., Murray, D., Orr, A., Porche, M., & Thompson, P. (2003).
17. Design considerations and field performance of an advanced automatic driller paper 79827.
18. Breiman, L., & Friedman, J. H. (1997). Predicting multivariate responses in multiple linear regression. Journal of Royal Statistical Society: Series B, 59(1), 3–54.
19. Brett, J., Warren, T. & Wait, D. (1990). Field experiences with computercontrolled drilling. SPE 20107.
20. Butler, N. A., & Denham, M. C. (2000). The peculiar shrinkage properties of partial least squares regression. Journal of Royal Statistical Society: Series B, 62(part 3), 585–593.
21. Carlson, J. M., & Doyle, J. (2002). Complexity and robustness. PNAS, 99(90001), 2538–2545.
22. Chen, H., & Allgower, F. (1998). A computationally attractive nonlinear ¨ predictive control scheme with guaranteed stability for stable systems. Journal of Process Control, 8(5–6), 475–485.

23. Choi, S. W., Yoo, C. K., & Lee, I. (2003). Overall statistical monitoring of static and dynamic patterns. Industrial and Engineering Chemistry Research, 42, 108–117.
24. Cohen, M. L. (1992). Epidemiology of drug-resistance—Implications for a postantimicrobial era. Science, 257(5073), 1050–1055.
25. Craig, W. A. (1998). Pharmacokinetic/pharmacodynamic parameters: rationale for antibacterial dosing of mice and men. Clinical Infectious Diseases, 26, 1–12.
26. Cunningham, R. (1978). An empirical approach for relating drilling parameters. Journal of Petroleum Technology, 30, 987–991.
27. Dewilde, P., & Deprettere, E. F. (1988). Singular value decomposition: An Introduction. In E. F., Deprettere (Ed.), SVD and signal processing: Algorithms, applications, and architectures (pp. 3–41). North-Holland.
28. Drlica, K. A. (2001). Strategy for fighting antibiotic resistance. ASM News, 67(1), 27–33.
29. Economides, M., Watters, L., & Dunn-Norman, S. (1988). Petroleum well construction. Wiley.
30. Eronini, I., Somerton, W., & Auslander, D. (1982). A dynamic model for rotary rock drilling. Journal of Energy Resources Technology, Transactions of the ASME, 104(2), 108–120.
31. Findeisen, R., & Allgower, F. (2001). The quasi-infinite horizon approach to ¨ nonlinear model predictive control. Nonlinear and adaptive control, Ncn4. Lecture Notes In Control and Information Sciences, 281, 89–108.
32. Frank, I. E., & Friedman, J. H. (1993). A statistical view of some chemometrics regression tools. Technometrics, 35(2), 109–148, with discussion.
33. Garthwaite, P. H. (1994). An interpretation of partial least squares. Journal of the American Statistical Association, 89(425), 122–127.
34. Gatlin, C. (1957). How rotary speed and bit weight affect rotary drilling rate. The Oil & Gas Journal, 55(20), 193–198.

35. Geladi, P. (1988). Notes on the history and nature of partial least squares (PLS) modeling. Journal of Chemometrics, 2, 231–246.
36. Geladi, P., & Kowalski, B. R. (1986). Partial least squares regression: A tutorial. Analytica Chimica Acta, 185, 1–17.
37. Geladi, P., Martens, H., Hadjiiski, L., & Hopke, P. (1996). A calibration tutorial for spectral data. Part 2. Partial least squares regression using matlab and some neural network results. Journal of Near Infrared Spectroscopy, 4, 243–255.
38. Giraldo, J., Vivas, N. M., Vila, E., & Badia, A. (2002). Assessing the (a)symmetry of concentration-effect curves: Empirical versus mechanistic models. Pharmacology & Therapeutics, 95, 21–45.
39. Gold, H. S., & Moellering, R. C. (1996). Antimicrobial-drug resistance. The New England Journal of Medicine, 335, 1444–1453.
40. Goutis, C. (1996). Partial least squares algorithm yields shrinkage estimators. Annals of Statistics, 24(2), 816–824.
41. Guo, R., & Sachs, E. (1993). Modeling, optimization and control of spatial uniformity in manufacturing processes. IEEE Transaction on Semiconductor Manufacturing, 6, 41–57.
42. Guo, X. (1988). A Preliminary investigation on the objective function of penetration cost applied to optimized drilling. SPE Drilling Engineering, 3(4), 411–418.
43. Ha, S., & Sachs, E. (1999). On-line control of process uniformity in single wafer processes. IEEE Transaction on Semiconductor Manufacturing, 12, 200–217.
44. Helland, I. S. (1990). Partial least squares regression and statistical models. Scandinavian Journal of Statistics, 17, 97–114.
45. Henson, M. (1998). Nonlinear model predictive control: current status and future directions. Computers & Chemical Engineering, 23(2), 187–202.
46. Holcomb, T. R., & Morari, M. (1992). PLS/Neural Networks. Computers & Chemical Engineering, 16(4), 393–411.

47. Holmes, P., Lumley, J. L., & Berkooz, G. (1996). Turbulence, coherent structures, dynamical systems, and symmetry. Cambridge University Press.
48. Hoskuldsson, A. (1988). PLS regression methods. ¨ Journal of Chemometrics, 2, 211–228.
49. Jackson, J. E. (1991). A user's guide to principal components. New York: Wiley Interscience. Jolliffe, I. T. (1986). Principal component analysis. Springer-Verlag.
50. Jorden, R., & Shirley, O. (1966). Application of drilling performance data to overpressure detection. Journal of Petroleum Technology, 1387–1394.
51. Jusko, W. (1971). Pharmacodynamics of chemotherapeutic effects: Dosetime-response relationships for phase-nonspecific agents. Journal of Pharmaceutical Sciences, 60, 892–895.
52. Kennedy, J. (1983). Fundamentals of drilling—Technology and economics. Tulsa, OK: PennWell Publishing.
53. Koshland, D. E. (1992). The microbial wars. Science, 257(5073), 1021.
54. Krischer, K., Rico-Martinez, R., Kevrekidis, I. G., Rotermund, H. H., Ertl, G., & Hudson, J. L. (1993). Model identification of a spatiotemporally varying catalytic reaction. AIChE Journal, 39, 89–98.
55. Kunin, C. M. (1993). Resistance to antimicrobial drugs—A worldwidecalamity. Annals of Internal Medicine, 118(7), 557–561.
56. Levy, S. B. (1994). Balancing the drug-resistance equation. Trends in Microbiology,2, 341–342.
57. Levy, S. B. (1998). The challenge of antibiotic resistance. Scientific American.
58. Lin, K., & Spanos, C. J. (1990). Statistical equipment modeling for VLSI manufacturing: An application for LPCVD. IEEE Transaction on Semiconductor Manufacturing, 3, 216–229.
59. Lipsitch, M., & Levin, B. R. (1997). The population dynamics of antimicrobial chemotherapy. Antimicrobial Agents and

Chemotherapy, 41(2), 363–373. Maidla, E. (2002). Personal communication.

60. Malthouse, E. C., Tamhane, A. C., & Mah, R. S. H. (1997). Nonlinear partial least squares. Computers & Chemical Engineering, 21, 875–890.

61. May, G. S., Huang, J., & Spanos, C. J. (1991). Statistical experimental design in plasma etch modeling. IEEE Transaction on Semiconductor Manufacturing, 4, 83–98.

62. Mayne, D., Rawlings, J., Rao, C., & Scokaert, P. (2000). Constrained model predictive control: Stability and optimality. Automatica, 26(6), 789–814.

63. McIntosh, A. R., Bookstein, F. L., Haxby, J. V., & Grady, C. L. (1996). Spatial pattern analysis of functional brain images using partial least squares. NeuroImage, 3, 143–157.

64. Misra, P. (1993). Studies on identification and nonlinearity assessment of multivariable control systems. Ph.D. thesis, Chemical Engineering Department, University of Houston.

65. Montgomery, D. C. (2001). Design and analysis of experiments. John Wiley and Sons.

66. Morari, M. (2002). Keynote speech at AIChE-CAST division award dinner. AIChE Annual Meeting.

67. Morari, M., & Lee, J. (1999). Model predictive control: Past, present, and future. Computers and Chemical Engineering, 23(4/5), 667–682.

68. Morens, D. M., Folkers, G. K., & Fauci, A. S. (2004). The challenge of emerging and re-emerging infectious diseases. Nature, 430, 242–249.

69. Mozumder, P. K., & Loewenstein, L. M. (1992). Method for semiconductor process optimization using functional representations of spatial variations and selectivity. IEEE Transactions on Components, Hybrids, and Manufacturing, 15, 311–316.

70. Mueller, M., de la Pena, A., & Derendorf, H. (2004). Issues in pharmacoki-˜netics and pharmacodynamics of anti-infective

agents: Kill curves versus MIC. Antimicrobial Agents and Chemotherapy, 48(2), 369–377.
71. National Committee for Clinical Laboratory Standards. (1997). Methods for dilution antimicrobial susceptibility tests for bacteria that grow aerobically. NCCLS Publication No. M7-A4. Villanova, PA.
72. National Institute of Allergy and Infectious Diseases. (2004). http://www.niaid.nih.gov/factsheets/antimicro.htm.
73. Neu, H. C. (1992). The crisis in antibiotic resistance. Science, 257, 1064–1073.
74. Nikolaou, M. (2001). Model predictive controllers: A critical synthesis of theory and industrial needs. In G. Stephanopoulos (Ed.), Advances in Chemical Engineering Series. Academic Press.
75. Nikolaou, M. (2003). A Challenge Problem in the Drilling Industry. AIChE Annual Meeting.
76. Nikolaou, M., & Bailey, A.D., III. (2002). Application of reduced-rank multivariate methods to the monitoring of spatial uniformity of wafer etching. Proceedings of MASM 2002.
77. Nikolaou, M., & Tam, V. H. (2005). A new modeling approach to the effect of antimicrobial agents on heterogeneous microbial populations. Journal of Mathematical Biology, in press.
78. Noble Corporation, 2005. http://www.noblened.com/Products/Overview.asp.
79. Ottino, J. M. (2003). Complex systems. AIChE Journal, 49(2), 292–299.
80. Papoulis, A. (1984). Probability, random variables, and stochastic processes. New York: McGraw-Hill.
81. Qin, S., & Badgwell, T. (1997). An overview of industrial model predictive control technology. In J. Kantor, C. Garcia, & B. Carnahan (Eds.), Fifth international conference on chemical process control—CPC V (pp. 232–256).

82. Qin, S., & Badgwell, T. (2000). An overview of nonlinear model predictive control applications. In F. Allgower & A. Zheng (Eds.), Nonlinear predictive control (pp. 369–393).
83. Qin, S. J., & McAvoy, T. J. (1992). Nonlinear PLS modeling using neural networks. Computers & Chemical Engineering, 16(4), 379–391.
84. Rao, C., Rawlings, J., & Lee, J. (2001). Constrained linear state estimation—a moving horizon approach. Automatica, 37(10), 1619–1628.
85. Rawlings, J. (2000). Tutorial overview of model predictive control. IEEE Contr. Svst. Magazine, 20(3), 38–52.
86. Rigopoulos, A., & Arkun, Y. (1996). Principal component analysis in estimation and control of paper machines. Computers & Chemical Engineering, 20, S1059–S1064.
87. Saputelli, L., Cherian, B., Gregoriadis, K., Nikolaou, M. Oudinot, C., Reddy, G., Economides, M., & Ehlig-Economides, C. (2000). Integration of computer-aided high-intensity design with reservoir exploitation of remote and offshore locations. SPE-64621.
88. Saputelli, L., Economides, M., Nikolaou, M., Demarchos, A., & Sagias, D. (2003). Real-time decision making for value creation while drilling and in well intervention, paper AADE-03-NTCE-12, AADE 2003 National Technology Conference. Practical Solutions for Drilling Challenges. Houston, TX.
89. Saputelli, L., Malki, H., Canelon, J., & Nikolaou, M. (2002). A critical overview of artificial neural network applications in the context of continuous oil field optimization. SPE-77703.
90. Savoji, M. H., & Burge, R. E. (1985). On different methods based on the Karhunen–Loeve expansion and used in image analysis. Computer Vision, Graphics, and Image Processing, 29, 259–269.
91. School Science, 2005. http://www.schoolscience.co.uk/content/4 /chemistry/ fossils/p6.html.

92. SIA, (2004). International Technology Roadmap for Semiconductors. http://www.itrs.net/Common/2004 Update/2004Update.htm.
93. Smith, T. H., Goodlin, B. E., & Boning, D. S. (1999). A Statistical analysis of single and multiple response surface modeling. IEEE Transactions on Semiconductor Manufacturing, 12, 419–430.
94. Spanos, P., Chevallier, A., Politis, N., & Payne, M. (2003). Oil and gas well drilling: A vibrations perspective. The Shock and Vibration Digest, 35, 85–103.
95. Stine, B. E., Boning, D. S., & Chung, J. E. (1997). Analysis and decomposition of spatial variation in integrated circuit processes and devices. IEEE Transactions on Semiconductor Manufacturing, 10, 24–41.
96. Tam, V., Schillling, A., Melnick, D., & Coyle, E. (2005). Comparison of beta-lactams in counter-selecting resistance of Pseudomonas aeruginosa. Diagnostic Microbiology and Infectious Disease, 52(2), 145–151.
97. Theodoridis, S., & Koutroumbas, K. (1999). Pattern recognition. Academic Press.
98. Turk, M., & Pentland, A. (1991). Eigenfaces for recognition. Journal of Cognitive Neuroscience, 3(1), 71–86.
99. Varaldo, P. E. (2002). Antimicrobial resistance and susceptibility testing: An evergreen topic. Journal of Antimicrobial Chemotherapy, 50, 1–4.
100. Vidyasagar, M. (2001). Randomized algorithms for robust controller synthesis using statistical learning theory. Automatica, 37(10), 1515– 1528.
101. Wagner, J. (1968). Kinetics of pharmacologic response. I. Proposed relationships between response and drug concentration in the intact animal and man. Journal of Theoretical Biology, 20, 173–201.
102. White, D. A., Boning, D. S., Butler, S. W. S. W., & Barna, G. G. (1997). Spatial characterization of wafer state using principal component analysis of optical emission spectra in plasma

etch. IEEE Transactions on Semiconductor Manufacturing, 10, 52–61.
103. Wise, R. (1998). Science, medicine, and the future—The development of new antimicrobial agents. British Medical Journal, 317(7159), 643–644.
104. Wold, H. (1982). Systems under Indirect Observation Using PLS. In C. Fornell (Ed.), A second generation of multivariate analysis: Methods: Vol. I. New York: Praeger. Wold, S. (1992). Nonlinear partial least squares modelling. II. Spline inner relation. Neural Computation, 1, 425–464.
105. Wold, S., Ruhe, A., Wold, H., & Dunn, D. J., III. (1984). The Colinearity problem in linear regression: The partial least squares (PLS) approach to generalized inverses. SIAM Journal on Scientific and Statistical Computing, 5(3), 735–743.
106. Yadav, P., (2002). Personal communication.
107. Yu, H. L., MacGregor, J. F., Haarsma, G., & Bourg, W. (2003). Digital imaging for online monitoring and control of industrial snack food processes. Industrial & Engineering Chemistry Research, 42(13), 3036–3044.

Chapter 7

Small-Hole Drilling in Engineering Plastics Sheet and its Accuracy Estimation

Hiroki Endo and Etsuo Marui

Department of Mechanical and Systems Engineering, Faculty of Engineering, Gifu University, 1-1 Yanagido, 501-1193 Gifu-shi, Japan

ABSTRACT

In recent manufacturing processes, the small diameter hole drilling process is frequently used owing to its good characteristics. The drilling process can easily be adapted to wide variations in lot size, processing accuracy, processing spot patterns where holes are made, and so on. Many machine elements, which have small diameter holes, are manufactured using engineering plastics of superior material and machining properties. However, it is not easy to drill holes with a diameter smaller than 1 mm, in recent machining technology as well. In this report, 1-mm diameter holes are drilled on two engineering plastics sheets and their drilling accuracy is discussed.

INTRODUCTION

Processing of small diameter holes is done in various materials, corresponding to the trend of downsizing or high accuracy in parts incorporated into electronic equipments, medical instruments or textile machineries. Many techniques are put to practical use, including drilling, ultrasonic machining, electric discharge machining, electrolytic machining, laser beam machining, electron beam machining, fluid or abrasive jet machining, and chemical blanking. Depending on the workpiece material, the machining accuracy, and the lot size, the best process for making holes of small diameter may be appropriately selected. Within these various machining processes, the drilling process can readily deal with a wide variety of machining conditions.

However, there are some difficult problems in drilling holes smaller than 1 mm in diameter. For example, a large load cannot be put on small drills, owing to their low strength and rigidity. Thus, the feed rate per unit drill rotation must be set small. The removal of drilled chips is difficult owing to the small drill flute area.

In many cases, engineering plastics are used in making various machine parts because they are light and have superior specific strength (that is, the ratio of tensile strength to density) compared with carbon steel. Also, the material cost of engineering plastics is competitive and their machinability is fairly good.

With these points as background, the orthogonal cutting of engineering plastics was investigated [1] and it was suggested here that the visco–elastic properties of engineering plastics have some effects on the magnitude of cutting force and the surface roughness of machined surfaces. There is a review paper [2] regarding the machining of engineering plastics. In this review paper, drilling process was also treated. It was pointed out that the heating up of the workpiece due to build-up of swarf on drill flutes is an obstacle to the drilling process of engineering plastics. Recently, some experiments have been attempted on drilling glass–fiber-reinforced engineering plastics sheets [3] and [4], and the thrust force and torque during drilling have been measured. In these papers, it was reported that the delamination phenomenon decreases the drilled hole integrity, when holes of about 5-mm diameter are drilled. However, the investigation on the accuracy in small hole drilling of engineering plastics is left pending.

Then in this paper, small diameter holes of 1 mm are drilled in two typical engineering plastics sheets, and the effect of spindle speed and feed rate on the accuracy (radius error) is estimated.

WORKPIECE MATERIALS

Two typical engineering plastics sheets, polyacetal (POM) and polyetherimide (PEI), were drilled. The materials properties are listed in Table 1.

Table 1: Material properties of workpiece engineering plastics

Performance	Unit	POM	PEI
Specific gravity		1.41	1.27
Rate of water	%	0.22	0.25
Melting point	°C	165	210
Coefficient of linear thermal expansion	cm/cm/°C	9×10^{-5}	5.6×10^{-5}
Tensile strength	MPa	61	124
Tensile extension (Yielding point)	%	40	23
Bending strength	MPa	89	157
Bending elasticity	GPa	2.60	3.07

Compressive strength	MPa	103	118
Izote impact value	J/m	74	42
Rockwell hardness	M scale	119	127

Polyacetal is a crystallized engineering plastics material. The main raw materials are acetal co-polymer and homo-polymer. POM has good fatigue properties and machinability. Many cams, guides and liners are made of POM. Very high accuracy is needed in these machined parts. PEI is an amorphous engineering plastic having superior thermal resistance characteristics. Special electrical parts, for example, electric insulators, connectors, are made of PEI, which is superior in mechanical strength but inferior in machinability to POM.

The workpiece size was: length 100 mm, width 50 mm and thickness 0.8 mm.

EXPERIMENTAL APPARATUS AND PROCEDURE

The drilling machine used is for small diameter holes, and is equipped with an automatic feed mechanism. A high-frequency induction motor positioned at the uppermost position of the main spindle drives the spindle. Maximum spindle speed is 12,500 rpm. The net spindle speed of the spindle during the drilling is measured by a tachometer, which counts number of the laser beam reflected from a reflective tape pasted on the scroll chuck.

A servomotor for drill feed drives the feed motion of the spindle. The feed is stepless, and a dial gauge equipped at the spindle head measures the length of the drill motion in the spindle axis direction. A stopwatch was used to measure the time needed for this length. The ratio of the moved length to the time is the substantial feed rate per unit time.

The spindle speed was varied between 1250 and 12,500 rpm. And also the feed rate per unit time was varied between 0.405 and

1.986 mm/s. Spindle speed was varied in keeping with the feed rate per unit time. Hence, the feed rate per unit drill rotation became small with the increase in the spindle speed of the drill.

The drill spindle end is attached to the scroll chuck. The drill used here is a conventional twist drill made of high-speed steel with a diameter of 1 mm. The various drill dimensions are as follows: flute length 17 mm, whole length 42 mm, point angle 124°, helix angle 30°, chisel edge length 0.28 mm, chisel edge angle 132°, lip height 0.008 mm. The scatterings of the point angle, the chisel edge length and the lip height are less than ±1°, ±0.01 mm and ±0.003 µm, respectively. There is no thinning of the chisel point. In some extra experiments, a 0.3 mm-diameter drill was also used. Such drills have no surface treatment. A dial gauge estimates deflection accuracy of the drill on the scroll chuck during rotation. Extreme care was taken so that the drill deflection was smaller than 5 µm.

The same drill made five holes under the cutting condition of the same spindle speed and the same feed rate. Another drill was used in the drilling under another cutting condition. Of course, the size accuracy of these drills exists within the above-mentioned size scattering. Any evidence of the wear of drills and the build-up of swarf on drill flutes were not recognized after five holes drilling.

Formerly mentioned workpiece of engineering plastics were set on the base of the drilling machine by clamping bolts. Dry cutting without fluid was performed.

CALCULATION OF DRILLED HOLE SHAPES

The 1-mm diameter holes drilled on engineering plastics sheets by the process described above are not geometrically true circles, but have a small radial deviation. Shape accuracy (radius accuracy) of the drilled holes is estimated by the following process.

An optical microscope equipped with digital measuring device measures the shape of the drilled hole. The cross wire of the microscope is set at the circumference of the hole. Then, the coordinates (x,y) of the hole circumference are read. Dividing the circumference into 18 equal parts, the same measurements are then repeated on each spot on the circumference. Using these 18 sets of measured qualities, the equation of the circle that fits closely to the drilled hole is calculated. This is called a least square circle, and in the calculation, the least squares method is applied.

The equation of the least square circle is assumed as follows:

$$x^2+y^2+Ax+By+C=0 \qquad (1)$$

Owing to the shape error of the hole, the right hand side of Eq. (1) does not become zero when the above-mentioned measured qualities (x_i, y_i) are substituted. The residual in this case is v_i and the following equation is obtained:

$$v_i = x_i^2 + y_i^2 + Ax_i + By_i + C \qquad (2)$$

Here, the coefficients A, B, C in Eq. (1) are determined as the sum of the squared values of the residual v_i becomes minimum. Values of these coefficients are obtained by solving the following simultaneous linear equations. In the calculation, N=18.

$$\begin{pmatrix} \sum_{i=1}^{18} x_i^2 & \sum_{i=1}^{18} x_i y_i & \sum_{i=1}^{18} x_i \\ \sum_{i=1}^{18} x_i y_i & \sum_{i=1}^{18} y_i^2 & \sum_{i=1}^{18} y_i \\ \sum_{i=1}^{18} x_i & \sum_{i=1}^{18} y_i & N \end{pmatrix} \begin{pmatrix} A \\ B \\ C \end{pmatrix} = - \begin{pmatrix} \sum_{i=1}^{18} x_i^3 + \sum_{i=1}^{18} x_i y_i^2 \\ \sum_{i=1}^{18} x_i^2 y_i + \sum_{i=1}^{18} y_i^3 \\ \sum_{i=1}^{18} x_i^2 + \sum_{i=1}^{18} y_i^2 \end{pmatrix} \qquad (3)$$

From this equation, coefficients A, B and C are obtained. Moreover, the coordinates (x_0, y_0) of the center of the least square circle and its radius r_m are obtained as follows:

$$x_0 = -\frac{A}{2} \qquad (4)$$

$$y_0 = -\frac{B}{2} \qquad (5)$$

$$r_m = \sqrt{\frac{A^2}{4} + \frac{B^2}{4} - C} \qquad (6)$$

Corresponding to the above process, the least square circles are described. An example is shown in Fig. 1, where the workpiece material is PEI, drill diameter: 1 mm, spindle speed: 12,500 rpm, and feed rate: 0.405 mm/s. The least square circle is indicated by the broken line.

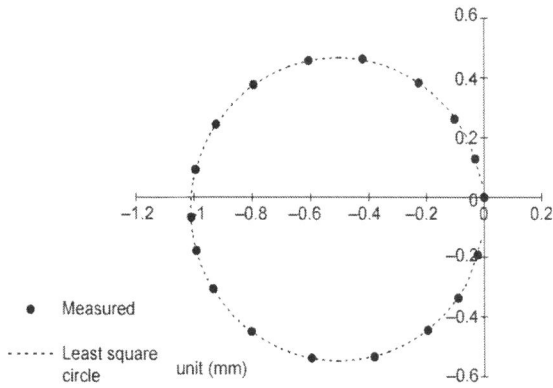

Figure 1: Example of least square circles.

ESTIMATION OF MACHINING ACCURACY AND EXPERIMENTAL RESULTS

Machining accuracy of the drilled holes is estimated by the radius error obtainable from the least square circles. The calculation process of the radius error is given here.

Radius r_i at the each measuring spot (x_i, y_i) is obtained from the coordinate of the least square circle center (x_0, y_0) of Eqs. (4) and (5) as follows:

$$r_i = \sqrt{(x_i - x_0)^2 + (y_i - y_0)^2} \tag{7}$$

Then, the radius error is calculated by the following equation. The parameter *rm* in the equation is the radius of the least square circle given by Eq. (6).

$$\Delta r_i = r_i - r_m \tag{8}$$

And the position of that measuring spot on the circle is represented by the following angle θ_i.

$$\theta_i = \tan^{-1}\left(\frac{y_i - y_0}{x_i - x_0}\right) \tag{9}$$

The relation between Δr_i and θ_i obtained from the above method is shown in Fig. 2 as a radius error curve. The drilling conditions in this figure are the same as those of Fig. 1 Three concavities and convexities are recognized on the circumference. Then, the drilled hole shape is approximately triangular. Similar results were obtained in other workpiece materials for other drilling conditions. Furthermore, it is seen that the circumference of the drilled hole exists in the vicinity within ±0.02 mm from the least square circle. This drilled hole shape is similar to that produced by the so-called drill walking phenomenon [5]. The radius of the least square circle is slightly larger than that of the drill. The difference between them is about 10 μm.

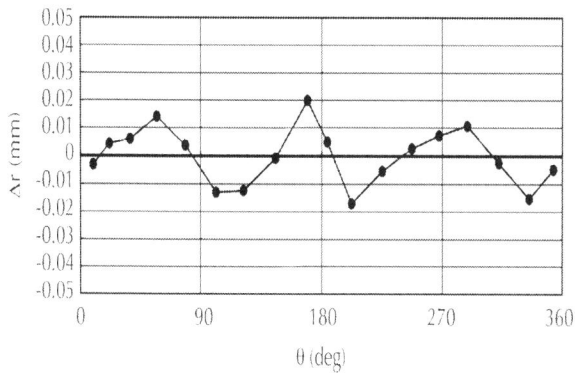

Figure 2: Example of machining error curves.

Result of Fig. 2 is obtained in the measurement at the drill entrance into workpiece. Small burr was formed at the drill exist and the accuracy measurement could not carry out as it is. Then, the burr was forcibly removed and the accuracy was measured. Almost the same accuracy was confirmed, because the workpiece is thin (0.8 mm thickness).

These radius errors are rearranged as functions of the spindle speed or the feed rate for every workpiece material. The results are given in Fig. 3, Fig. 4, Fig. 5 and Fig. 6. Error bars indicate the distribution range of the experimental data. The radius error becomes small hyperbolically with the increase in the feed rate and becomes large linearly with the spindle speed. Small diameter drills were used in this experiment and their bending rigidity is low. Rotational cutting speed is almost zero near the chisel point. At that point, the drill has only a small axial velocity corresponding to the drill feed motion. Accordingly, the rate of penetration [6] is extremely small when the feed rate is small. As mentioned above, the walking phenomenon occurs owing to small errors in drill size. This phenomenon is compounded with the effect of small rate of penetration when small feed rate and large spindle speed are applied. Hence, the positioning accuracy of the drill point against the workpiece is not very good at small feed rate and large spindle speed. As a result, it is supposed that the radius error becomes large. For example, the rate of penetration, that is the feed rate per unit drill rotation, is about 2 μm when the drill rotation speed is 12,500 rpm and the feed rate is 0.405 mm/s. The rate of penetration is about 100 μm when the drill spindle speed is smallest (1250 rpm) and the feed rate is largest (1.986 mm/s).

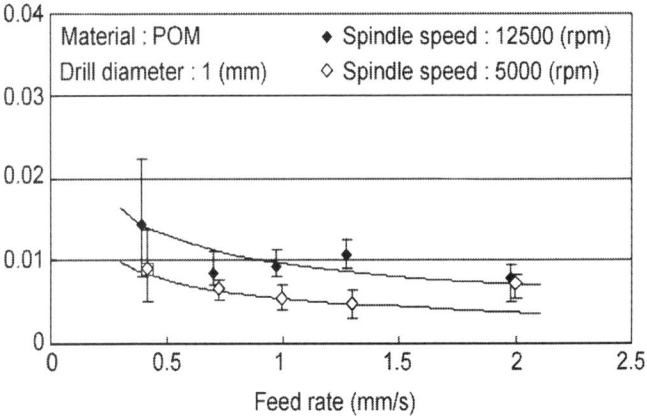

Figure 3: Effect of feed rate on radius error (POM).

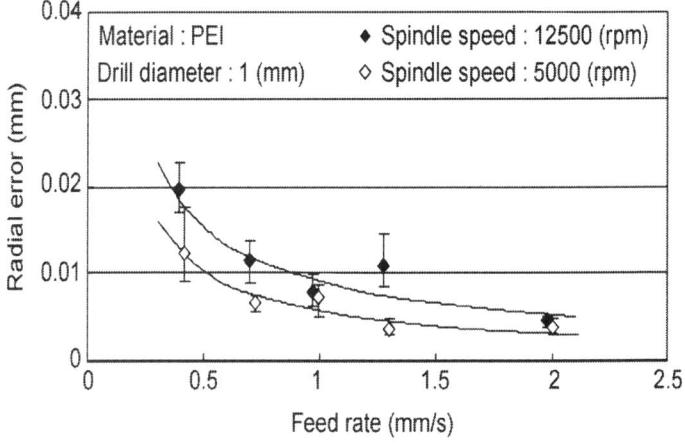

Figure 4: Effect of feed rate on radius error (PEI).

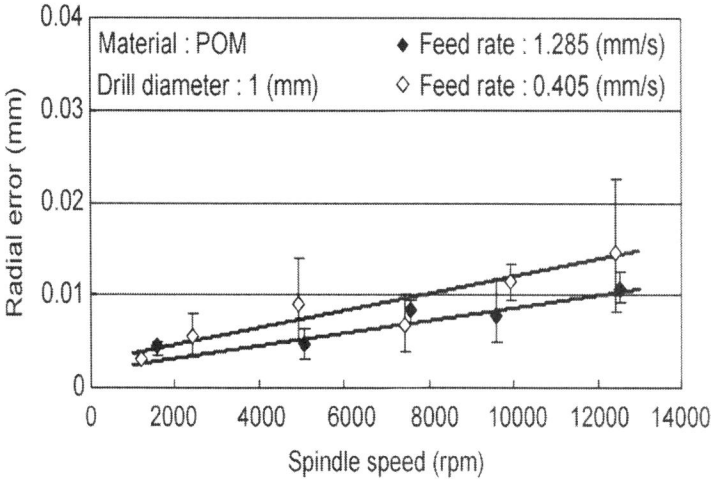

Figure 5: Effect of spindle speed on radius error (POM).

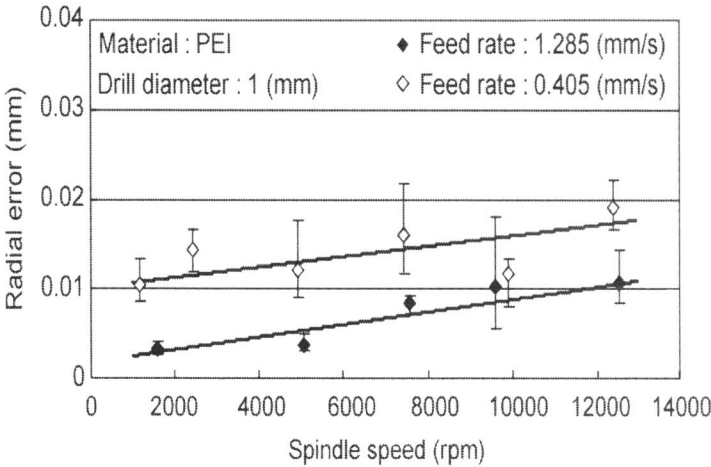

Figure 6: Effect of spindle speed on radius error (PEI).

One reason for the radius error worsening when the rotation speed becomes high is that chatter [6] related to the drill dynamic characteristics is possible. However, the small drill size errors and the relative drop in the feed rate per unit drill rotation corresponding to the spindle speed increase have a large effect on the radius error. In conclusion, it is important to drill a small hole in the drilling condition so as to maintain a sufficiently high feed rate per unit drill rotation.

An example of superposition of the results of POM and PEI is given in Fig. 7. It is recognized in this figure that the radius accuracy in the drilling of PEI is slightly inferior to that of POM. PEI is a kind of supper engineering plastics. PEI is superior to POM in tensile strength, compressive strength, bending strength, bending elasticity and Rockwell hardness, as seen in Table 1. Owing to this, the machinability of PEI may be worse than that of POM and the result of Fig. 7 regarding the radius error was be obtained.

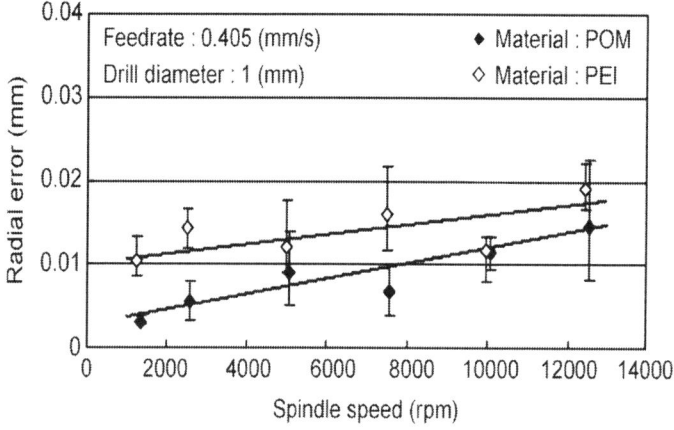

Figure 7: Effect of spindle speed on radius error (POM and PEI, feed rate 0.405 mm/s).

DRILLING EXAMPLE OF 0.3 MM DIAMETER HOLES

By way of a trial, drilling was performed using a 0.3 mm diameter drill under the same drilling conditions.

Examples of the least square circle and the radius error curve are shown in Fig. 8 (POM) and in Fig. 9 (PEI). The spindle speed and the feed rate in this drilling are the largest within the overall drilling conditions. We can safely drill holes of 0.3 mm diameter in both POM and PEI workpieces. The radius of the least square circles is slightly larger than that of the drill. The difference between them is about 3 μm. The radius error is smaller than ±3 μm, so the drilled hole accuracy is fairly good.

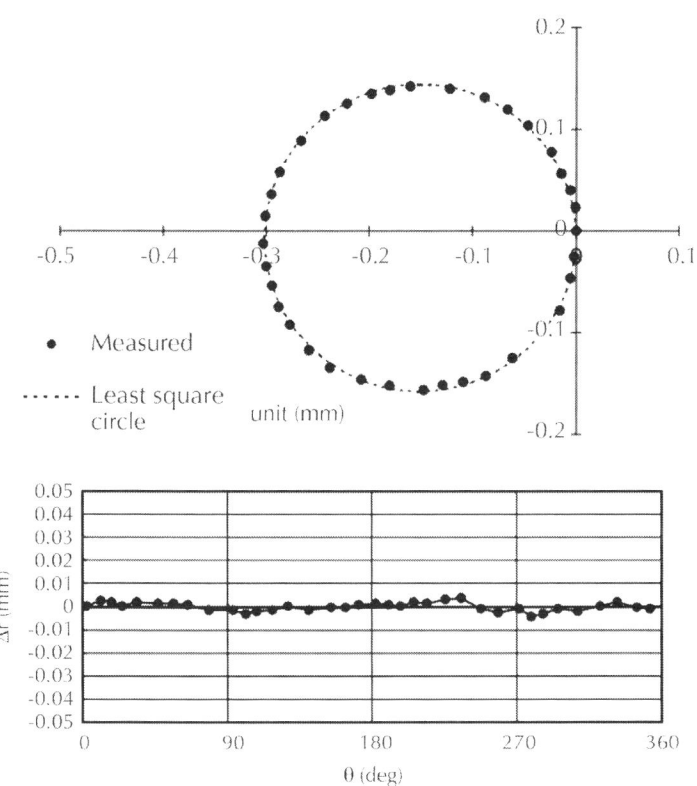

Figure 8: Least square circle and radius error curve (POM, drill diameter 0.3 mm, spindle speed 12,500 rpm, feed rate 1.986 mm/s).

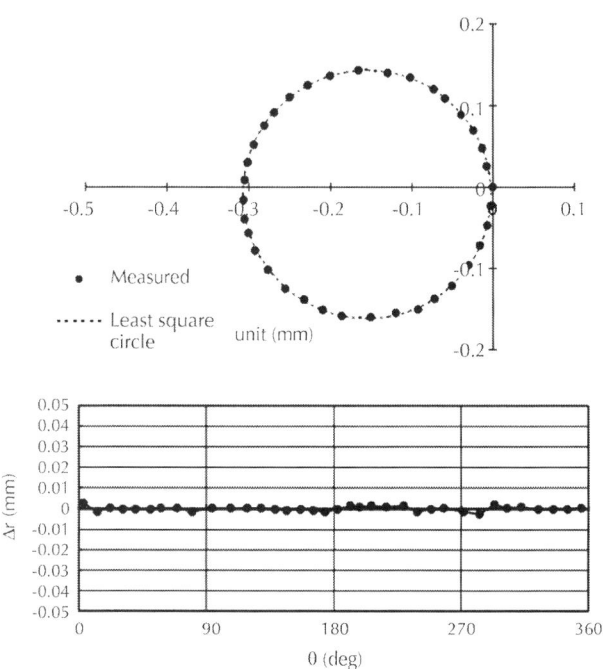

Figure 9: Least square circle and radius error curve (PEI, drill diameter 0.3 mm, spindle speed 12,500 rpm, feed rate 1.986 mm/s).

CONCLUDING REMARKS

Small holes were drilled in two engineering plastics sheets POM and PEI using a drill 1 mm in diameter. Drilling can be done on both workpiece materials.

Reading the drilled hole shape by optical micrometer, and calculating the least square circle, the drilling accuracy (radius error) can be estimated. The radius error becomes worse when the drill feed rate is small and the spindle speed is large. The feed rate per unit drill rotation (rate of penetration) is relatively small when the spindle speed is large. Hence, it is supposed that the positioning accuracy of the drill against the workpiece is not good, and that the radius error becomes worse under the drilling conditions in which the feed rate per unit drill rotation is small. From this fact, it is desirable that small diameter holes be drilled in the condition in which the feed rate does not become low.

Drilling of POM and PEI sheets using 0.3 mm diameter drill was undertaken. The radius error in this drilling is smaller than ±3 μm. Fairly accurate holes are obtained.

REFERENCES

1. K.Q. Xiao, L.C. Zhang, The role of viscous deformation in the machining of polymers, International Journal of Mechanical Science 44 b (2002) 2317–2336.
2. M. Alauddin, I.A. El Baradie, M.S.J. Hashmi, Plastics and their machining: a review, Journal of Materials Processing Technology 54 (1995) 40–60.
3. W.-C. Chen, Some experimental investigations in the drilling of carbon fiber-reinforced plastic (CFRP) composite laminates, International Journal of Machine Tools and Manufacture 37 (1997) 1097–1108.
4. E. Capello, Workpiece damping and its effect on delamination damage in drilling thin composite laminates, Journal of Materials Processing Technology 148 (2004) 186–195.
5. M. Tsueda, Y. Hasegawa, H. Kimura, On walking phenomenon of drill, Transactions of the JSME 27 (1961) 816–823.
6. D.F. Galloway, Some experiments on the influence of various factors on drill performance, Transactions of the ASME 79 (1957) 191–231.

Chapter 8

Study on the Diamond Tool Drilling of Engineering Ceramics

Q.H. Zhang[a], J.H. Zhang[a], D.M. Sun[b], and G.D. Wang[c]

[a]College of Mechanical Engineering, Shandong University, No. 73, Jingshi Road, Jinan 250061, PR China
[b]College of Material Engineering, Shandong University, Jinan, PR China
[c]Shandong Machine Design and Research Institute, Jinan, PR China

ABSTRACT

A method for drilling holes in engineering ceramics by using a diamond tool has been developed. In this method, a drilling tool rotates with fixed abrasives. The machining mechanism of drilling based on the fracture mechanics concept is analyzed, and a new theoretical model of the material removal rate is proposed. According to this model, the material removal rate increases in accordance with the increase of the static load applied, the rotational speed of the drilling tool, and the grain size of the abrasive. Selecting 99.5% Al_2O_3 ceramics as the workpiece material, experiments have been carried out. The results show that diamond drilling is an effective method for machining engineering ceramics.

INTRODUCTION

Engineering ceramics have numerous excellent physical and mechanical properties: high hardness, high thermal resistance, chemical stability, and low thermal and electrical conductivity, to name but a few. Because of these special qualities, engineering ceramics are expected to be used increasingly in a number of high-performance applications ranging from electronic and optical devices to heat- and wear-resistant parts [1], [2] and [3]. Until today, their applications have mostly been limited to electronic and optical devices. One reason is to be found in the limitation on the forming process prior to sintering, which restricts the generation of complex geometry and makes it difficult to ensure adequate accuracy and surface finish. There is also a considerable deficit in terms of the production or machining of more complex geometries in the hardened post-sintering state, with limitations on either the performance or the forming capacity of the majority of the processes in current use. Machining engineering ceramics to final dimensions by conventional methods is extremely laborious and time consuming. Tight tolerances and dimensions with acceptable surface and sub-surface damage are something only attainable at great cost. Thus research into the areas of more efficient material removal processes have been beginning to gather momentum in recent years, especially in the ways and means of reducing the occurrence of faults or cracks in the sub-surface of the machined ceramics [4], [5] and [6].

A kind of machining method for drilling holes in engineering ceramics by using a rotary diamond tool is proposed in this paper. It can increase the material removal rate, and improve the surface finish.

This paper intends to further the understanding of the basic mechanisms the diamond tool drilling of ceramics and thus to enable the prediction of the material removal rate in terms of the static load applied, the grain size of the abrasive, and the rotational speed of the drilling tool.

THE MECHANISM OF DIAMOND TOOL DRILLING

The process of diamond tool drilling is shown schematically in Fig. 1. The tool is rotating in the drilling process, the workpiece is stationary. The material removal mechanism has been investigated generally by microscopic observation of the abraded surface. The process of material removal is thought to be similar to that of a single tool cutting in that in all cases material is removed by an individual tool or particle displacing or fracture the work surface. The workpiece material is found to be 'stabbed' off in the form of many minute particles by the abrasive grains grinding.

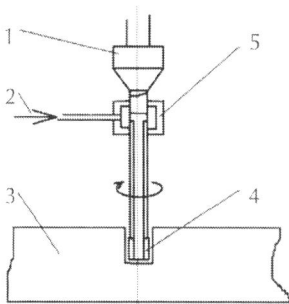

Figure 1: Schematic diagram of the diamond tool drilling process: (1) chuck, (2) water, (3) workpiece, (4) tool, (5) water jacket.

It is concluded that the removal of engineering ceramic occurs primarily by brittle fracture in diamond tool drilling. To understand this process, it is helpful to study the indentation of brittle materials, because the abrasive grains acting on the workpiece surface are just like indenters.

Investigation of Indentation in Ceramics

The deformation and fracture pattern observed under the normal contact of ceramics by a Vickers indenter is illustrated in Fig. 2. Directly under the indenter is a zone of plastic deformation. Two principal crack systems have been identified, which emanate from the plastic zone: median/radial cracks and lateral cracks. The behavior of both types of cracks is effected by residual stresses from the non-uniform plastic deformation in the elastic/plastic material. Radial cracks are initiated by a wedge-like action during loading, and they may continue to propagate during unloading due to residual tensile stresses acting on the crack-tip. Lateral cracks are observed to initiate and propagate by residual stresses only as the indenting load is removed. The initiation and propagation of radial as well as lateral cracks are considered to contribute greatly to the material removal process. As shown in Fig. 2, the initiation and propagation of these radial and lateral cracks at the end lead to chipping of the brittle material. A critical load P_c for initiating a radial crack is given by [7]:

$$P_c = \alpha \frac{K_{IC}^4}{H_V^3} \tag{1}$$

where α is a dimensionless factor related to the indenter geometry, K_{IC} is the fracture toughness of workpiece, and H_V is the Vickers hardness of the workpiece material.

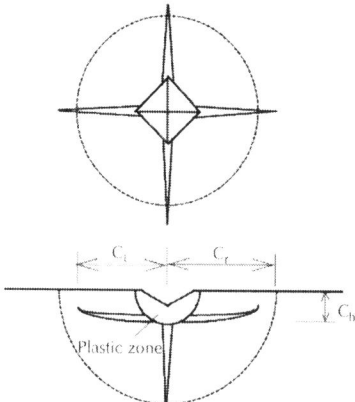

Figure 2: Localized deformation and fracture of ceramics due to indentation.

For the size of the median or radial crack C_r and lateral crack C_L, respectively, the following equations have been derived [8]:

$$C_r = \xi_1 P^{1/2}(H_V^{1/4} K_{IC}^{1/3}) \qquad C_L = \xi_2 \left(\frac{P}{K_{IC}}\right)^{3/4} \qquad (2)$$

where P is the load applied, and ξ_1 and ξ_2 are proportional constants.

It is generally regarded that the depth of the lateral crack C_h is proportional to $(P/H_V)^{1/2}$:

$$C_h = \xi_3 \left(\frac{P}{H_V}\right)^{1/2} \qquad (3)$$

where ξ_3 is a proportionality constant [9].

It is concluded from these results that the size of the median/radial or lateral crack grows with an increase in the load and with a decrease in the fracture toughness of the workpiece material. Investigations on indentation described so far provide useful information for understanding the practical diamond drilling process of ceramics.

The Relationship between the Material Removal Rate and Various Parameters

According to the test results of ceramics in indentation, a model of material removal caused by a single abrasive is proposed, as shown in Fig. 3. The model takes into account the linear tangential motion of the abrasive along the surface. Assuming that an individual abrasive grain follows a linear path with a constant depth of cut, the volume of workpiece material removed by an abrasive grain (indenter) under a normal load P is proportional to the dimensions of the lateral crack and the length of travel d. The volume of removed workpiece material V_0 for one grain is obtained as (see Fig. 4):

$$V_0 = 2C_L C_h d \tag{4}$$

where C_L is the length of the lateral crack, C_h is the depth of the lateral crack, d is the acting distance of the grain.

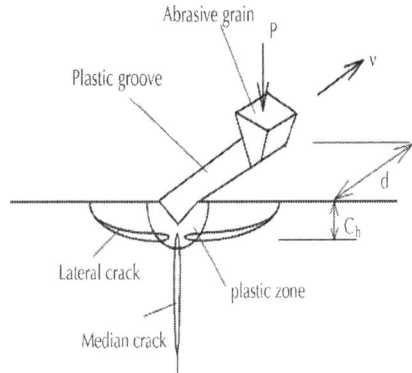

Figure 3: Schematic model of chip formation.

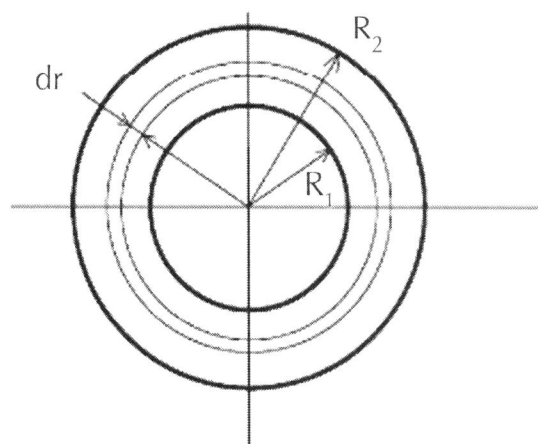

Figure 4: Schematic diagram of the terminal face of the drilling tool.

Then, the material removal rate for one grain is given by:

$$M_{V_0} = 4\pi C_L C_h \omega r \qquad (5)$$

where ω is the rotational speed of the tool, and r is the radius of the grain's track.

Assuming that the density of the effective cutting grains is λ, the number of effective grains in area dA (seeFig. 4) is:

$$n = \lambda\, dA = 2\lambda \pi r\, dr \qquad (6)$$

Then, the material removal rate of the tool is:

$$M_V = \int_{R_1}^{R_2} 8\pi^2 \lambda C_L C_h \omega r^2\, dr = \frac{8}{3}\pi^2 \lambda C_L C_h \omega (R_2^3 - R_1^3) \qquad (7)$$

The number of effective grains in the terminal face of the tool is [10]:

$$N = \lambda A = \frac{K_1}{d_0^2}\left(\frac{6\upsilon_g}{\pi}\right)^{2/3} \pi(R_2^2 - R_1^2) \qquad (8)$$

where $A = \varpi(R_2^2 - R_1^2)$, R_2 is the external radius of the diamond drilling tool, R_1 is the internal radius of the diamond drilling tool, K_1 is proportionality constant; υ_g is the concentration of abrasive grains, and d_0 the mean diameter of the abrasive grains. Then:

$$\lambda = \frac{K_1}{d_0^2}\left(\frac{6\upsilon_g}{\pi}\right)^{2/3} \qquad (9)$$

The load acting on a single abrasive grain is:

$$P = \frac{W}{N} = W\left(\frac{d_0^2}{\pi K_1}\right)\left(\frac{6\upsilon_g}{\pi}\right)^{-2/3} (R_2^2 - R_1^2)^{-1} \qquad (10)$$

Substituting (2),(3) ,(9), (10) and into Eq. (7):

$$M_V = \frac{8}{3}\pi^{3/4}\xi_1\xi_2\omega \left(\frac{d_0^2}{K_1}\right)^{1/4}\left(\frac{6\upsilon_g}{\pi}\right)^{-1/6}$$
$$\times K_{IC}^{-3/4} H_V^{-1/2} W^{5/4} \frac{R_2^3 - R_1^3}{(R_2^2 - R_1^2)^{5/4}} \qquad (11)$$

Eq. (11) can be simplified to:

$$M_V = K d_0^{1/2} \upsilon_g^{-1/6} K_{IC}^{-3/4} H_V^{-1/2} \omega W^{5/4} \qquad (12)$$

where K is a proportional constant.

According to Eq. (12), the material removal rate M_V will be increased with the increase of the applied static load W, the rotational speed of the tool ω, and the size of the abrasive grains d_0.

EXPERIMENTAL PROCEDURE

Experiments were performed on a drilling machine. The diamond drilling tool, which was especially designed to accept coolant (see Fig. 1), was a special diamond wheel with a external radius of $R_2=10$ mm and an internal radius of $R_1=6$ mm. Three types of diamond drilling tool were prepared, their grits being 80, 120, and 160, and their concentrations were being 100%.

A 99.5% Al_2O_3 ceramic was selected as the workpiece material and water was selected as the coolant.

The MRR is measured through a dial gauge with an accuracy of 0.001 mm. The depth of drilling per minute can be measured with a dial gauge, and then the MRR can be calculated (The MRR is the cross-sectional area of the drilling tool multiplied by the depth of drilling per minute.). The surface roughness is measured using a Talysurf 4' (England) surface measuring instrument with a relative accuracy of 5%.

EXPERIMENTAL RESULTS AND DISCUSSION

The Effect of the Static Load

Test results show that the material removal rate tends to increase with the increase of the static load, as shown in Fig. 5, which is similar to Eq. (12).

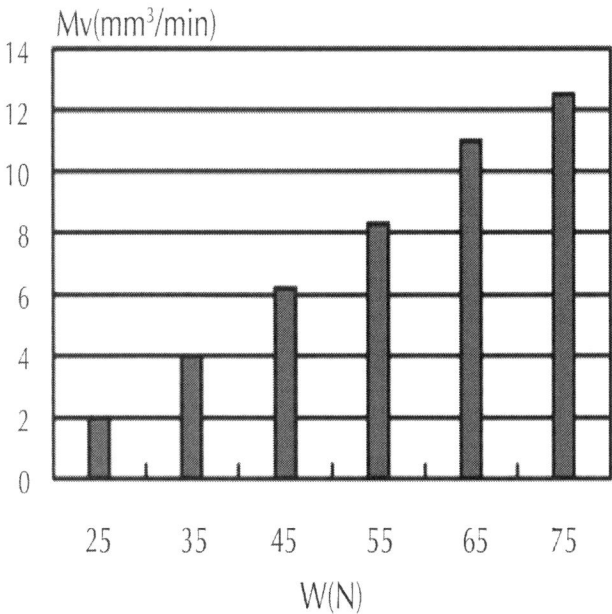

Figure 5: The effect of the static load.

The surface roughness is found to be slightly affected by the applied static load, increasing with the increase of the static load. The present experiments were conducted under the following conditions: =720 rev/min, C=80. When the static load W is 25, 35, 45, 55, 65 N, the surface roughness (Ra) is 0.0040, 0.0040, 0.0042, 0.0045, and 0.0045 mm, respectively.

The Effect of the Rotational Speed of the Drilling Tool

Fig. 6 shows the effect of the rotational speed of the drilling tool on the MRR. An increase of the rotational speed of the drilling tool causes an increase in the MRR. As there are some other factors affecting the MRR, for example the flushing of swarf and the self-sharpening of abrasive grains, the MRR is not proportional to the rotational speed of the drilling tool.

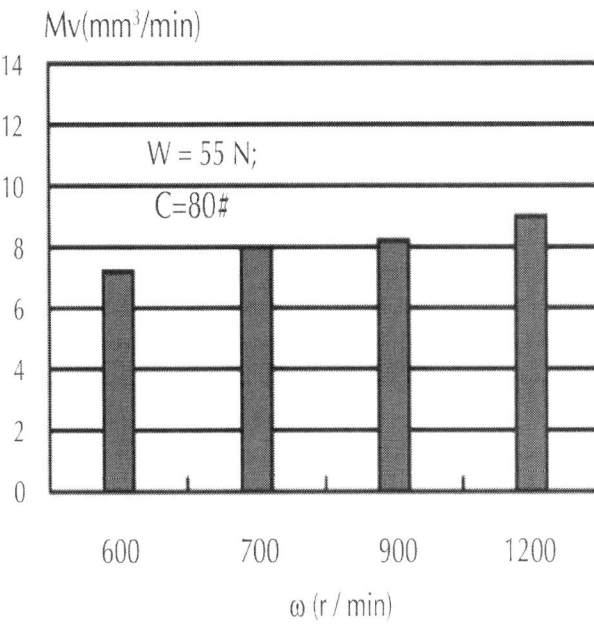

Figure 6: The effect of the rotational speed of the drilling tool.

It is very important for high material removal rate to flush away the swarf. With the increase of the rotational speed of the drilling tool, a lot of swarf is formed, and flushing it away becomes increasingly more difficult. This will also affect the self-sharpening of the abrasive grains. Generally speaking, self-sharpening of abrasive grains include two components, self-sharpening through progressive abrasive grain fragmentation and self-sharpening through progressive bond erosion. When the swarf is not completely flushed away, bond erosion and abrasive grain fragmentation become difficult, resulting in the dulling of the drilling tool: thus the MRR will be affected. Additionally, the flushing of swarf is related to the pressure of the coolant. In the present experiments, the pressure of the coolant is 2.0×10^4 Pa. Under this condition, loading occurred when the rotational speed of the drilling tool was 1200 rpm.

The surface roughness decreases with an increase of the rotational speed of the drilling tool. The present experiments were conducted under the following conditions: W=55 N, C=80. When the rotation speed ω is 520, 720, 900 and 1120 rev/min, the surface roughness (Ra) is 0.0055, 0.0045, 0.0042, and 0.0042 mm, respectively.

The Effect of the Grain Size

According to Eq. (12), the material removal rate will increase with the increase of the grain size, and this is confirmed by the test results, as shown in Fig. 7.

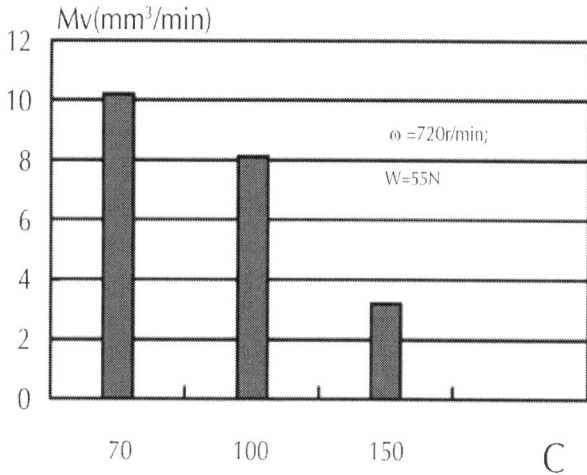

Figure 7: The effect of grain size.

The size of the abrasive grain can greatly affect the surface roughness. The surface roughness depends mainly on the size of the abrasive grains. The greater is the grain size, the rougher is the finished surface. The present experiments were conducted under the following conditions: W=55 N, =720 rev/min. When the C is 80, 120, and 160, the surface roughness (Ra) is 0.0045, 0.0040, and 0.0020 mm, respectively.

CONCLUSIONS

The basic mechanism of the diamond tool drilling of ceramics has been studied, and the effect on the material removal rate has been explored. The test results show that any increase in terms of the static load applied, the rotational speed of the drilling tool, and the size of abrasive grains, results in an increase of the material removal rate. The relationship between the roughness of the finished surface and each parameter is given: the finished surface roughness increase with increase of the static load, decrease of the rotational speed, and increase of the abrasive grain size.

REFERENCES

1. R.W. Davidge, Mechanical Behavior of Ceramics, Cambridge University Press, Cambridge, 1979.
2. T. Warren Liao, Flexural strength of creep feed ground ceramics: general pattern, ductile–brittle transition and MLP modeling, Int. J. Mach. Tools Manuf. 38 (4) (1998) 257–275.
3. S. Reschke, C. Bogdanow, Engineering ceramics: new perspectives through value-added (multi-) functionality, Key Eng. Mater. 175–176 (1999) 1–10.
4. K. Suzuki, T. Uematsu, S. Mishiro, A new grinding method for ceramics using a biaxially vibrated nonrotational ultrasonic tool, Ann. CIRP 42 (1) (1993) 375–378.
5. K.P. Rajurkar, Z.Y. Wang, A. Kuppattan, Micro removal of ceramic material (Al2O3) in the precision ultrasonic machining, Precis. Eng. 23 (1999) 73–78.
6. K. Ueda, T. Sugita, H. Hiraga, A J-integral approach to material removal mechanisms in microcutting of ceramics, Ann. CIRP 40 (1) (1991) 61–64. Fig. 6. The effect of the rotational speed of the drilling tool. Fig. 7. The effect of grain size. Q.H. Zhang et al. / Journal of Materials Processing Technology 122 (2002) 232–236 235

7. A.G. Evans, D.B. Marshall, in: D.A. Rigney (Ed.), Fundamentals of Friction and Wear of Materials, ASME, New York, 1981, pp. 439–442.
8. I.A. Markov, Machining of Intractable Materials with Ultrasonic and Sonic Vibrations, Illife Books, 1966.
9. M. Komaraiah, P.N. Reddy, A study on the influence of workpiece properties in ultrasonic machining, Int. J. Mach. Tools Manuf. 33 (3) (1993) 495–505.
10. U. Kuniaki, E. Kazuhito, The trial construction of ultrasonic grinding equipment and the machining characteristics, JSPE 52 (1) (1986) 107–113.

Chapter 9

Mathematical Model of Dissolution of Particles of NaCl in Well Drilling: Determination of Mass Transfer Convective Coefficient

L.A. Calcada[a], L.A.A. Martins[a], C.M. Scheid[a], S.C. Magalhães[a], and A.L. Martins[b]

[a]Department of Chemical Engineering, Federal Rural University of Rio de Janeiro, BR-467, Km 7, Campus da UFRRJ, 23, 890-000 Seropédica, Rio de Janeiro, Brazil

[b]PETROBRAS S.A./CENPES/PDP/TEP, Av. Hum Quadra 07, Ilha do Fundão, 21494-900 Rio de Janeiro, RJ, Brazil

ABSTRACT

The drilling of salt formations generates particulate material composed of salt mixtures. In water-based muds (WBM) along the annular section, these salt mixtures tend to dissolve. The dissolution can cause changes to the drilling fluids. Its can lead to the well enlarging, to salt gravels accumulating at the bottom of the borehole, and other operational problems. The objective of this work is to study the dissolution of salt particles in brine and evaluate the mass transfer coefficient. The experimental data of the salt dissolved in brine along a flow system was obtained through an experimental apparatus consisting essentially of a mixer tank and a flow line. This study proposes a model based on the mass conservation equations for brine and solid salt particles. Samples of brine were taken along the flow system with the aim of evaluating the profiles of brine concentration versus position. The experiments were conducted for different operational conditions of brine flow and solid particles fed into the system. The experimental data were used as input for evaluating the overall coefficient of mass transfer. The system of partial differential equations (PDE's) was discretized in space by finite differences. The discretized model becomes a system of ordinary differential equation (ODE's) that can be solved using a subroutine LSODE in FORTRAN language. The hypothesis of steady state simplifies the model, allowing it to be solved analytically. The mathematical model is able to simulate the experimental data under different conditions, showing that the proposed model is capable of predicting experimental outputs under the studied range. The average deviation between the experimental values and the simulated values was less than 2.3%.

INTRODUCTION

To exploit the oil reservoirs of the pre-salt layer, one must overcome great technological challenges. Thus in Brazil, where there are five main offshore producing basins, demand is a growing for studies on drilling in this layer. The most promising and numerous basins are considered to be the relatively unexplored producing basins of Campos, Espírito Santo, and Santos, São Paulo. In addition, experts believe that waiting to be found are a number of giant discoveries (Santos et al., 2009).

Drilling in salt layers is difficult because of the complexity of the rocks' saline formations where new technologies are in demand. Among the problems commonly encountered when drilling these formations, one of the most challenging is the dissolution of salt cuttings in water-based mud (WBM). These cuttings, generated by the drill bit, enter the annular region and tend to dissolve naturally in water-based mud (WBM) during its upward return flow to the surface. These dissolved cuttings can produce changes in rheology and other physicochemical properties of the drilling fluid. They can also give rise to several problems during the drilling, including the accumulation of salt cuttings at the bottom of the well, the expanding and weakening of the borehole walls and, even, the collapse of these walls (Durie and Jessen, 1964 and Aksel'rud et al., 1992).

One way to minimize the effects of salt dissolution is to use synthetic fluid. Its application, however, is rather limited due to its high cost, impact on the environment, and its tendency to complicate the evaluation of formations. Many researchers have tried exploring this field using WBM. One concern with this is salt particles dissolving during the flow of the fluid within the well. A good method to minimize this problem is to use saturated water, although it is difficult to control the rheological properties of this type of fluid.

The literature presents many works focused in the phenomenon of leaching salt caverns (Aksel'rud et al., 1992). The aim of this work, though, is to study the dissolution of salt particles during the flow of suspension of particles of NaCl in brine. In this context, a fluid flow apparatus was constructed. We obtained experimental data of the brine concentration and salt particle dissolution varying with position. These data were used to evaluate the overall convective mass transfer coefficient. The phenomenological equations were obtained based on conservation of mass for the system composed of brine phase and solid salt particles. Those equations generate a system of Partial Differential Equations (PDE's). Two solutions were proposed to solve the model, a numerical solution for transient state and an analytical solution for the steady state. The numerical solution was obtained using an algorithm in FORTRAN applying finite differences method to discretize the geometrical coordinate. Parameter estimation was performed by the method of maximum likelihood. A functional dependence was proposed between the overall mass transfer coefficient and concentration of the brine.

In this work, for the composition of salt in the drilling fluids we use real conditions to obtain a brine solution commonly used that can be in the range of seawater concentration (32 g/L) to near saturation values (360 g/L). It was desirable to explore a wide range of initial brine concentrations (C_f) in the experiments and simulations, and variables such as fluid flow and the rate of salt particles fed in the flow.

LITERATURE REVIEW

Drilling Fluids

Drilling fluids are designed for the drilling process to serve numerous functions: to clean the rock fragments generated from the action of the drill bit and carry them to the surface, exert sufficient hydrostatic pressure to prevent formation damage (fractures) and the flow of undesired fluids from the formation into the well, as well as to cool and lubricate the bit. Also, the drilling fluid should not damage the formation or cause difficulties to the use of formation evaluation techniques (Darley and Gray, 1988, Bourgoyne et al., 1991, Caenn and Chillingar, 1996, Luckham and Rossi, 1999 and Menezes et al., 2010). Some properties of a drilling fluid have to be controlled. Many additives are used to control density (bentonite, barite, carbonate, etc.), rheological behavior (carboxymethylcellulose, xanthan gum, etc.), pH (NaOH, $NaCO_3$), etc. There are two primary types of drilling fluids (muds), oil-based muds (or synthetic – OBM) and water-based muds (WBM). In offshore operations, WBMs are formulated using sea water. They are easy to handle, can be disposed of in the ocean, and their preparation relies on a near and free source. On the other hand, in the drilling of the salt layer WBMs dissolve the salt particles and the wall of the well. The use of saturated WBMs is very tricky to be used as an effective way to prevent physical–chemical and rheological changes during the process. To control rheological properties many additives are generally used as polymers, such as xanthan gum (Benchabane and Bekkour, 2006, Hamed and Belhadri, 2009, Hamida et al., 2009 and Benyounes et al., 2010) and sodium carboxymethylcellulose (Benchabane and Bekkour, 2006,Iscan and Kok, 2007, Benyounes et al., 2010 and Menezes et al., 2010). The xanthan gum interacts with dissolved salt and changes the rheological behavior of WBMs.

OBMs, which are usually inert, have excellent properties of stability (Meng et al., 2012). Using OBMs could minimize the effects of salt dissolution. The literature reports that the most effective drilling fluids are based on oil – crude or synthetic. Their application is limited, though, due to their high costs and their harsh impact on the environment (Caenn and Chillingar, 1996, Barret and Eugene, 2005, Hamed and Belhadri, 2009,Meng et al., 2012 and Sönmez et al., 2013). Synthetic fluids are included in the oil-based fluid category. Common ones are based on esters, ethers, polyalphaolefins, glycols, glycerines and glucosides (Caenn and Chillingar, 1996). More expensive than WBMs, these muds have been developed to share similarities with WBMs, including stability, ease-of-use, and a soft impact on the environment. Where synthetic fluids are commonly used is in the drilling of wells of high instability owing to salt dissolution. In the drilling of salt formations, the muds used most often are synthetic fluids.

Kinetics of Dissolution

Aksel'rud et al. (1992) obtained the coefficient of mass transfer for salt particle in a fluidized bed. Formed by pressing finely ground powders of salt, the salt particles used were cylindrical with 9 mm diameters. Eq.(1) was the one adopted by the authors. Parameter K evaluated for NaCl was 1.0×10^{-4} m/s.

$$-\frac{dm_{salt}}{dt} = Ka(C-C^*). \tag{1}$$

Using the same concept as the mass transfer coefficient, Morse and Arvidson (2002) studied the dissolution of calcite minerals from the Earth's surface. The authors studied the dissolution and deposition of calcite in the formation of geological layers. They could model that dissolution using Eq. (2)

$$-\frac{dm_{calcite}}{dt} = \frac{A}{V}K(1-\Omega)^n. \tag{2}$$

Finneran and Morse (2009) presented a study on the kinetics of dissolution of calcite in saline waters based on the model

$$R = K(1-\Omega_{calcite})^n. \tag{3}$$

They investigated the effects of the ionic strength and the temperature on the dissolution of calcite. Solutions of KCl and NaCl were investigated. It was found that a first order kinetic was sufficient to describe the rate of dissolution and the parameter K was dependent of the solution's composition.

Alkattan et al. (1997) studied the kinetics of halide dissolutions. The dissolution rates of compressed halite powders were measured using rotating disk techniques at constant halite saturation states and in the presence of trace concentrations of aqueous F^-, Br^-, and I^-. Where they considered a mass model according to the coefficient of each ion as

$$\frac{dm_{Na^+}}{dt} = \frac{dm_{Cl^-}}{dt} = K(C-C^*). \tag{4}$$

Those equations used in the models described have some similarities. In fact, they came from the classical literature but each equation was adapted to its need. Some of the classical equations used on dissolution kinetics can be seen on Bird et al. (2002).

MATERIAL AND METHODS

Flow Line

An experimental unit was built to obtain the profile of the brine concentration as a function of time and position due to the dissolution of particles of salt at 30 °C. The flow line was composed of a 2000 l mixing tank that was attached to a mechanical 1.5 hp stirrer and a 3 hp helical displacement positive pump. The mixer tank fed the brine solution into a PVC flow line gutter that had a diameter of 150 mm, a length of 29 m, and an inclination of 5°. It was here the dissolution process occurred. The salt particles were fed into the gutters at the inlet using a solid feeder. At the outlet of the flow line there was a 500 L receiving tank. Pumping the fluid to a 1500 L storage tank was a $\frac{3}{4}$ hp centrifugal pump. The apparatus also had strainers, which gathered salt particles before the concentration flowed into the receiving tank. Fig. 1 shows the layout of the unit.

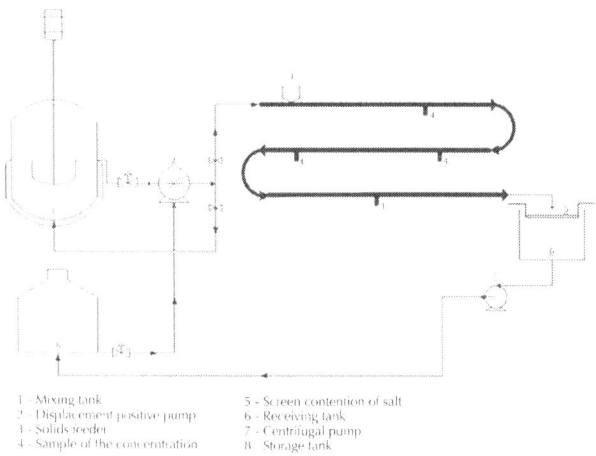

Figure 1: Schematic of the experimental unit for disposal of salts.

The brine samples were taken from the gutters and drained simultaneously to a reservoir after the system had sufficient time to reach a steady state. The points of sample collection were located at 5.5, 13.0, 20.5, and 26.7 m from the solid feeder. At the end of the flow line, the remaining solids were collected on a screen and dried. Knowing the mass of the remaining solids it was possible to determine the mass of the dissolved salt by a simple mass balance.

An experimental grid was made using a 2 (number of conditions proposed for each variable)×2 (number of independent variables) factorial design with a central point to each of the four initial brine concentrations (C_f) (see Table 1). For C_f higher than 90 g/L, the experimental grid was changed to improve the experimental analysis. In those cases, the value Q was maintained at 2 L/s and W was set to 4 distinct values.

Table 1: Table of experimental values at 30 °C

Experiments	C_f (g/L)	Q (L/s)	W (g/s)
1	32.0	1.00	24
2	32.0	1.00	48
3	32.0	2.00	24
4	32.0	2.00	48
5	32.0	1.50	36
6	90	1.00	24
7	90	1.00	48
8	90	2.00	24
9	90	2.00	48
10	90	1.50	36
11	175	2.00	40
12	175	2.00	30
13	175	2.00	35
14	175	2.00	20

15	258	2.00	40
16	258	2.00	30
17	258	2.00	35
18	258	2.00	20

The experiments with values at the central point (Experiments 5 and 10) and the experiments with 35 g/L of the variable W (Experiments 13 and 17) were used to validate the model. In the initial concentrations of 32 and 90 g/L, the value of the minimum flow rate was enough to be able to suspend all salt particles without any accumulation in the tubes. Nonetheless, to prevent salt deposition during the flow, the study set the value of Q to 2 L/s in the experiments with 175 and 258 g/L of C_f.

Mass Conservation Equations

The mathematical model for the dissolution of NaCl particles in the flow of brine consists of two partial differential equations based on the conservation of mass in the liquid and solid phases. The hypotheses considered in those equations were as follows: (i) Transient state, plenty turbulent flow, isothermal and incompressible fluid, (ii) unidirectional flow along the gutters and (iii) two-phase flow (liquid phase-solution of water and NaCl (brine) and solid phase-particles of solid NaCl).

The resulting equations of mass balance for the salt solution and mass balance for salt in the solid phase become respectively Eqs. (6) and (7).

Mass balance for brine

$$\frac{\partial}{\partial t}C(z,t) + \bar{v}_z \left(\frac{\partial}{\partial z} C(z,t) \right) = ka(C^* - C(z,t)),$$

I.C., $C(z, 0) = C_f$,

B.C., $C(0, t) = C_f$. \hfill (6)

Mass balance for the particles of salt

$$\rho_s\left(\frac{\partial}{\partial t}\varepsilon_s(z,t)+\bar{v}_z\left(\frac{\partial}{\partial z}\varepsilon_s(z,t)\right)\right)=-ka(C^*-C(z,t)),$$

I.C., $\varepsilon_s(z,0)=\varepsilon_{s0}$.

B.C., $\varepsilon_s(0,t)=\varepsilon_{s0}$. (7)

The first term in both equations is related to time in a control volume, the accumulation term. The second term in both equations is the convective mass transfer term due to the flow velocity. The third term in both equations is related to the mass transfer between the liquid and solid phases.

The interface area "a" is the total surface area for mass transfer per unit of volume, which is represented by Eq. (8). This equation is adapted from McCabe et al. (1985).

$$a=\frac{6\varepsilon_s(z,t)}{\bar{D}_p},$$ (8)

where $\bar{D}_p$ is the Sauter mean diameter, defined as Brennen (2005). The system of partial differential equations described above can be solved by numerical methods using finite differences and an algorithm that was implemented in FORTRAN language.

$$\bar{D}_p=\frac{1}{\int_0^1 \frac{dX}{D_p}}\simeq\frac{1}{\sum_i \frac{\Delta X_i}{D_{pi}}}.$$ (9)

Analytical Solution

Although the system of differential partial equations (Eqs. (6) and (7)) cannot be solved analytically, what is possible is an analytical solution for the condition of steady state. Eqs. (6) and (7) become:

Mass balance for brine (steady state)

$$\bar{v}_z\left(\frac{\partial}{\partial z}C(z,t)\right)=Ka(C^*-C(z,t)),$$

B.C., $C(0)=C_f$ (10)

Mass balance for the salt particles (stead state)

$$\overline{v}_z \left(\frac{\partial}{\partial z} \varepsilon_s(z,t) \right) = \frac{-Ka(C^* - C(z,t))}{\rho_{sal}},$$

B.C., $\varepsilon_s(0) = \varepsilon_{s0}$ (11)

The equation of continuity for the whole system can be written as

$$\frac{\partial C}{\partial z} + \rho_s \frac{\partial \varepsilon_s}{\partial z} = 0.$$ (12)

In this case, the analytical solution can be compared with the experimental data since the experiments were also done in steady state. The analytical solutions for the equations above are shown in Eq. (13).

$$C = \frac{C^* - a_2 a_3 e^{a_1 Z}}{1 - a_2 e^{a_1 Z}},$$ (13)

where the constants a_1–a_3 are given by Eqs. (14), (15) and (16) respectively.

$$a_1 = \frac{6K(C^* - C_0 - \rho_s \varepsilon_0)}{v_z \rho_s D_p},$$ (14)

$$a_2 = \frac{(C^* - C_0)}{\varepsilon_0 \rho_s}.$$ (15)

$$a_3 = C_0 + \rho_s \varepsilon_0.$$ (16)

RESULTS AND DISCUSSIONS

Experimental and Mathematical Analysis

Initially, in order to evaluate uncertainty, the five experiments of initial concentration of brine (C_i) of 32 g/L and the experiments

10, 13, and 17 were done in triplicate. Table 2 shows the mean value of concentration and the average deviation for each of the five experiments of $C_f = 32 g/L$.

Table 2: Results for concentration in function of position to experiments in initial brine concentration of 32 g/L and the average deviation

Position (m)	Concentration $\delta = \pm 2.5$ (g/L)				
	Exp. 1	Exp. 2	Exp. 3	Exp. 4	Exp. 5
5.5	49.0	65.6	39.6	43.7	46.0
13.0	53.2	69.6	42.3	50.1	50.6
20.5	53.9	70.6	42.7	51.2	50.9
26.7	54.6	71.4	43.9	53.3	51.4

The triplicates of the experiments 10, 13, and 17 show similar deviations. The average uncertainty of those experiments were 5%, 2.5%, 1%, 0.5% for the experiments of initial brine concentrations of 32 g/L, 90 g/L, 175 g/L and 258 g/L, respectively.

The Convective Mass Transfer Coefficient

In our experiments, the global mass transfer coefficient was affected neither by the Reynolds number nor the temperature. Indeed, the temperature was stable at 30 °C and the volumetric flow of brine was fully turbulent. In the parameter evaluation, the global mass transfer coefficient showed that it was dependent on the concentration of the brine. We evaluated several models to fit K with brine concentrations and Eq.(17) was the one that gave the better results as

$$K = A1 C^{A2} \quad (17)$$

That estimation was a result of using the method of maximum likelihood. That method tends to estimate the parameter "K" with the smallest error between the simulated curves and the experimental data (Anderson et al., 1978). The simulated curves generated by the model for several initial brine concentrations are presented in Fig. 2, Fig. 3, Fig. 4 and Fig. 5. Unfortunately, in some experiments, data overlapped for a given condition of fluid and solid flow rate generating data in the same region. The same happened for simulated data.

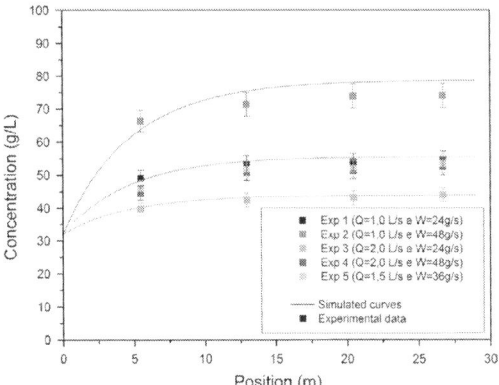

Figure 2: Brine concentration in function of position. Experiment and model data in initial brine concentration of 32 g/L.

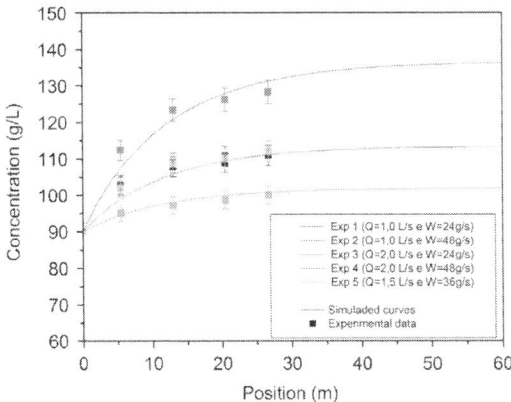

Figure 3: Brine concentration in function of position. Experiment and model data in initial brine concentration of 90 g/L.

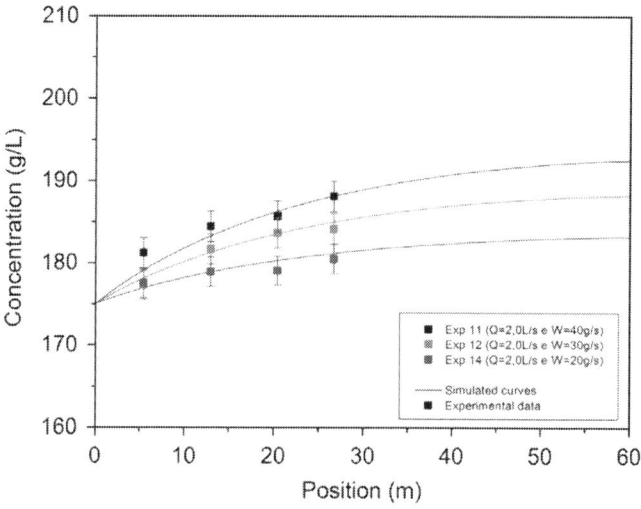

Figure 4: Brine concentration in function of position. Experiment and model data in initial brine concentration of 175 g/L.

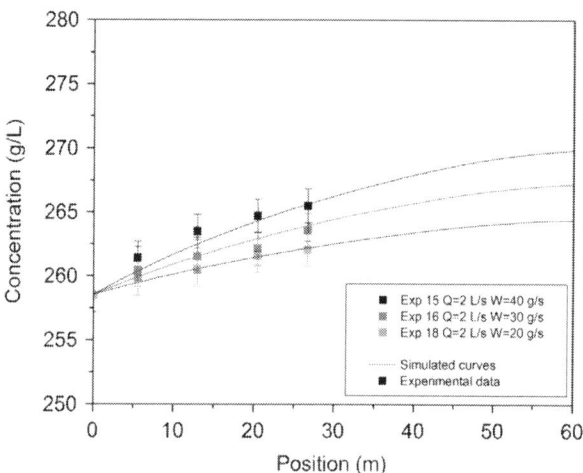

Figure 5: Brine concentration in function of position. Experiment and model data in initial brine concentration of 258 g/L.

Note that, the results presented in that figure are related to the experiments used in the parameter evaluation. Experiments 5,

10, 13, and 17 were used only to validate the model. The final expression for K becomes Eq. (18) where the parameters A_1 and A_2 were found to be 0.00489 and −0.639.

$$k = 0.00489(C_f)^{-0.639} \qquad (18)$$

Fig. 6 shows the global mass transfer coefficient as a function of brine concentration fitted by the model. We observed that the evaluated parameter was in a range of 2–6×10^{-4} m/s. It was observed that the value presented in Eq. (18) was greater than 1.0×10^{-4} m/s found by Aksel'rud et al. (1992). This discrepancy between values may be due to the different operating conditions adopted by the authors. Aksel'rud et al. (1992) offered no Reynolds number range to which their coefficient was estimated.

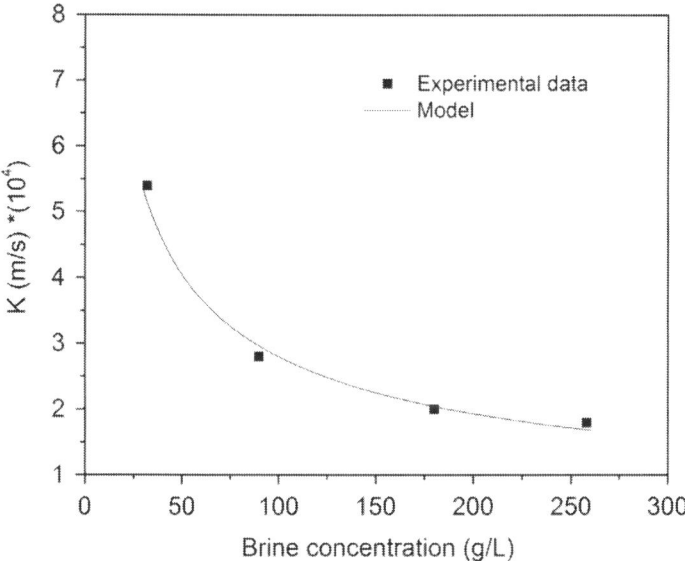

Figure 6: The global mass transfer coefficient as a function of brine concentration.

We also generated simulated curves using the analytical solution presented by Eq. (13), valid for steady state. The results are presented by Fig. 7, Fig. 8, Fig. 9 and Fig. 10. The experimental conditions of data presented in Fig. 7, Fig. 8, Fig. 9 and Fig. 10 are the same presented in Table 1. The results for the transient and steady state were very similar since this transient process is very fast. We started to collect the experimental data after 60 s. Those figures show graphically when the system reached steady state, that is, when the numerical solution (function of time) had reached the analytical solution (steady state). For the analytical solution, we used a constant value for the k coefficient calculated by Eq. (18) using the value of the concentration of brine in the beginning of the experiment.

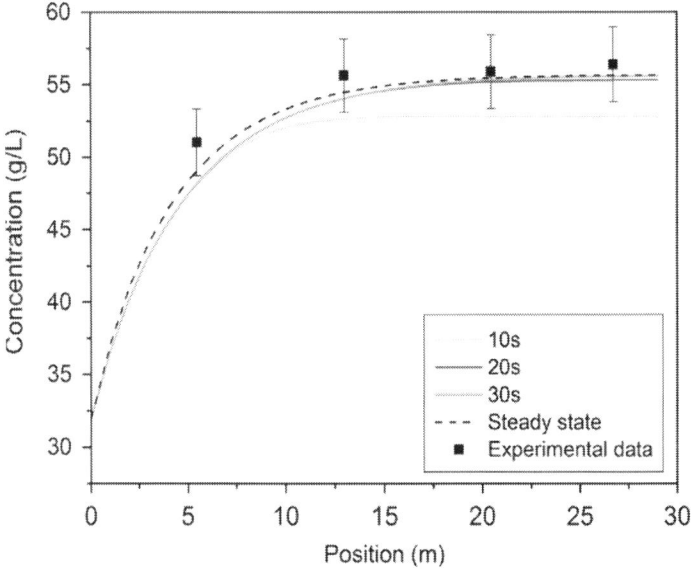

Figure 7: Brine concentration in function of position. Experiment and model (numerical and analytical) data in initial brine concentration of 32 g/L (Exp. 5).

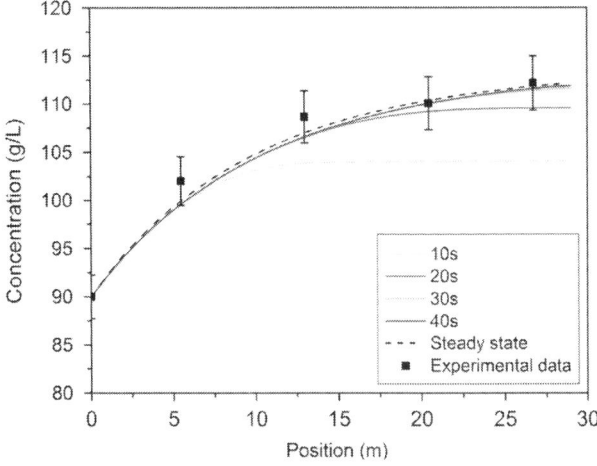

Figure 8: Brine concentration in function of position. Experiment and model (numerical and analytical) data in initial brine concentration of 90 g/L (Exp.10).

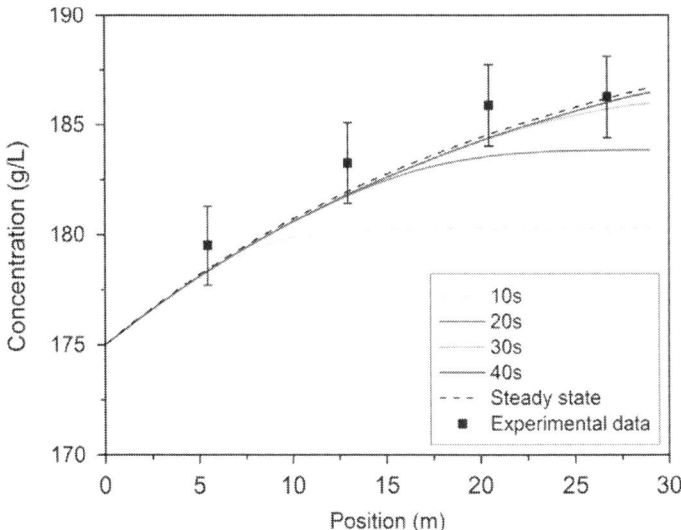

Figure 9: Brine concentration in function of position. Experiment and model (numerical and analytical) data in initial brine concentration of 175 g/L (Exp. 13).

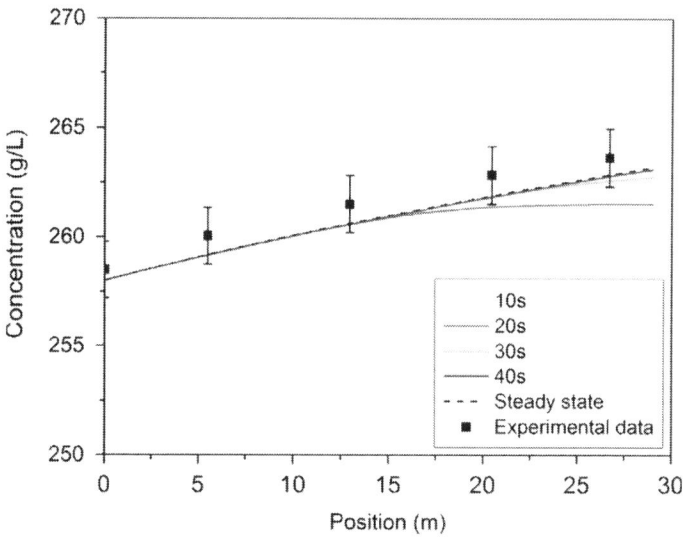

Figure 10: Brine concentration in function of position. Experiment and model (numerical and analytical) data in initial brine concentration of 258 g/L (Exp. 17).

Fig. 7, Fig. 8, Fig. 9 and Fig. 10 show that the simulated curves fit the experimental data. Note that those experimental data were not used on the parameter evaluation. Also, the model was able to predict profiles of concentration in different conditions from those used as input data for the estimation. The solution curves for different values of time showed how fast the system reached a steady state. In all cases, the simulated curves showed that the steady state was close to 40 s. The samples of concentration in the experiments in steady state were taken later, near 60 s. The average error between the simulated curves for steady state and the validation experimental data was less than 2.3%. Thus, the analytical solution showed itself to be appropriate for generating simulations to steady state. The numerical solution was a more complex simulation method and it generated solutions to a transition state.

Fig. 11, Fig. 12, Fig. 13 and Fig. 14 show the profiles of volume fraction of particles in function of position for the experiments with initial brine concentration of 32 g/L, 90 g/L, 175 g/L, and 258 g/L.

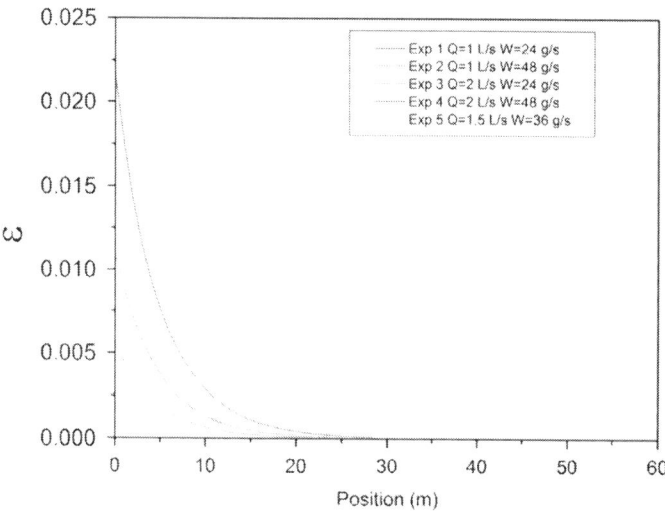

Figure 11: Volume fraction of particles in function of position. Model data in initial brine concentration of 32.0 g/L.

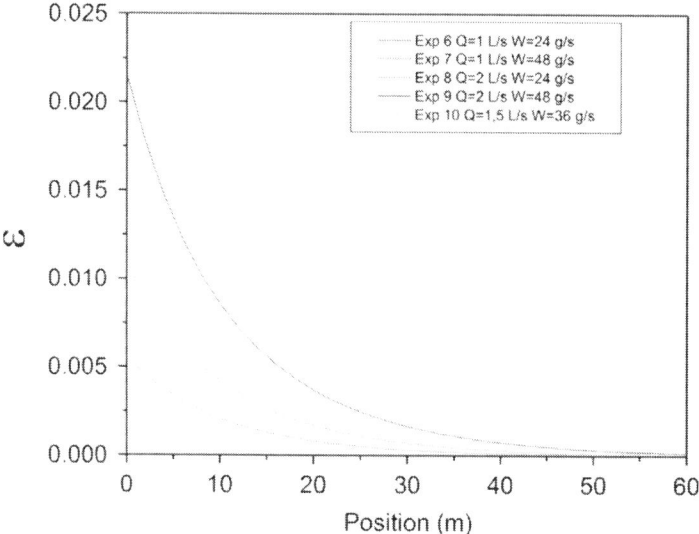

Figure 12: Volume fraction of particles in function of position. Model data in initial brine concentration of 90.0 g/L.

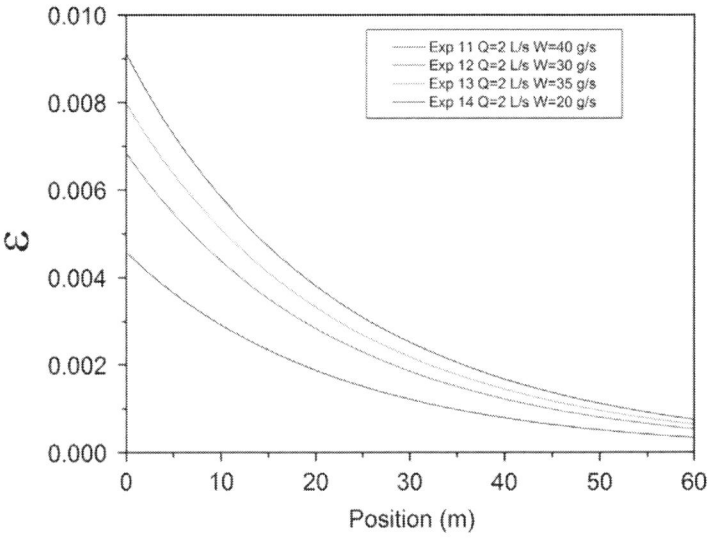

Figure 13: Volume fraction of particles in function of position. Model data in initial brine concentration of 175.0 g/L.

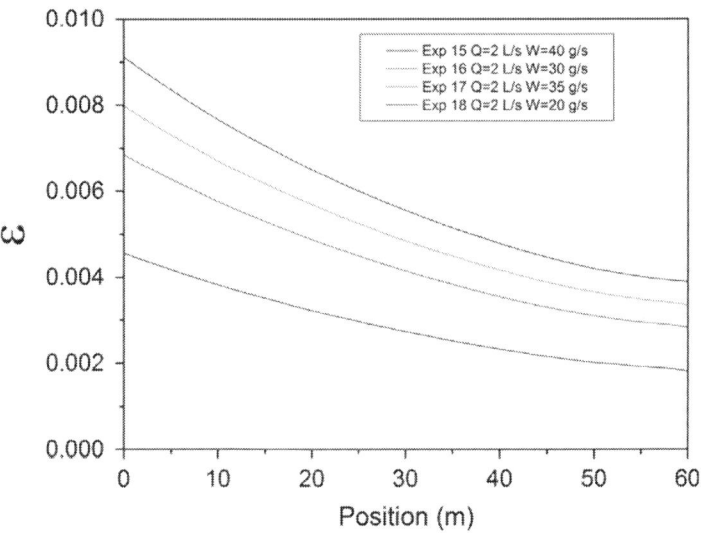

Figure 14: Volume fraction of particles in function of position. Model data in initial brine concentration of 258.0 g/L.

The behavior of salt dissolution in function of position can be observed in the figures above. At the lowest concentration (32 g/L), the volume fraction of solids decreases, in the first few meters, markedly fast. This is due to the greater potential for dissolution of diluted brine solutions. The amount of salt decreases rapidly, stabilizing up to 15 m, where almost all salt had already dissolved. From this point, so small is the quantity of salt available for dissolution that the concentration of brine stabilizes.

At higher concentrations (90 g/L, 175 g/L and 258 g/L), the dissolution occurred at a slower rate, as the concentration decreases the kinetics of dissolution.

It is observed that, with the exception of the simulations of concentration of 32 g/L, the simulations for the concentration of brine as the volume fraction for the limit of the simulation occur up to 60 m. The experimental unit has 29 m, and the last measuring point is approximately 26 m, but these simulations were made for the 60 m limit so one can visualize better the dynamics of different experiments at different concentrations.

At a concentration of 32 g/L, it can be observed in both simulations (concentrations and volume fraction) that the dissolution occurs mainly in the first 15 m. After this, dissolution becomes very limited due to the small concentration of salt present in the particulate system. Up to 90 g/L simulated curves show that there are still salt particulates at the very end of the pipeline and the brine concentration is still growing. In the simulations of 178 g/L and 258 g/L, it is not possible to observe, even up to 60 m, when the amount of salt particulate comes close to zero. Thus, it can be stated that the concentration of the solution directly affects the kinetics of dissolution. Hence the concentration of the liquid phase increases the distance required for dissolving a given amount of salt in the flow. In some cases, some simulated curves overlapped as we can see in Fig. 11 and Fig. 12.

CONCLUSIONS

We have presented here our construction of an experimental fluid flow line and a mathematical model to enable a study of the dissolution of salt particles. It was observed that an increase in the brine concentration decreased the kinetics of dissolution. Thus, solutions with higher concentrations were less able to dissolve. We noted that on experiments with initial concentrations of brine of 32 g/L, the dissolution of most all the salt occurred within the first 15 m. In higher concentrations, most of the particles dissolved up to advanced positions. In addition, the parameter K is dependent of the NaCl concentration. Increasing the concentration of the brine lowered the dissolution kinetics. The mathematical model fitted the data of brine concentration and volume fraction of solids. The average percentage error between values generated by the model and the experimental data for these experiments was less than 2.3%. We showed that the global mass transfer coefficient was in a range of $2-6 \times 10^{-4}$ m/s. The mathematical model can be helpful in simulations of the dissolution of particles of salt in brine.

ACKNOWLEDGMENTS

The authors give thanks for the financial support offered by CENPES (PETROBRAS Research Center)(4600293210) and cooperation in scientific support from CAPES and PPGEQ/UFRRJ.

REFERENCES

1. Anderson, T.F., Abrams, D.S., Grens II, E.A., 1978. Evalutaion of parameters for nonlinear thermodynamic models. AIChE J. 24 (1), 20–29.
2. Aksel'rud, G.A., Boiko, A.E., Kashcheev, A.E., 1992. Kinetics of the solution of mineral salts suspended in a liquid flow. J. Eng. Phys. 61 (1), 885–888.

3. Alkattan, M., Oelkers, E.H., Dandurand, J.L., Schott, J., 1997. Experimental studies of halite dissolution kinetics. Chem. Geol. 137, 201–219.
4. Barrett, B., Bouse, E., 2005. Handbook of Drilling Fluids. Elsevier Inc., United States of America.
5. Benchabane, A., Bekkour, K., 2006. Effects of anionic additives on the rheological behavior of aqueous calcium montmorillonite suspensions. Rheol. Acta 45, 425–434.
6. Benyounes, K., Mellak, A., Benchabane, A., 2010. The effect of carboxymethylcellulose and xanthan on the rheology of bentonite suspensions. Energy Sources Part A 32, 1634–1643.
7. Bird, R.B., Stewart, W.E., Lightfoot, E.N., 2002. Transport phenomena, Chemical Engineering Department, Second Ed. University of Wisconsin-Madison, United States of America.
8. Bourgoyne Jr., A.T., Millheim, K.K., Chenevert, M.E., Young Jr, F.S., 1991. Applied Drilling Engineering, Second Printing Society of Petroleum Engineers, Richardson, TX.
9. Brennen, C.E., 2005. Fundamentals of Multiphase Flow. Cambridge University Press, United States of America.
10. Caenn, R., Chillingar, G.V., 1996. Drilling fluids: state of the art. J. Pet. Sci. Eng. 14, 221–230.
11. Darley, H.C.H., Gray, G.R., 1988. Composition and Properties of Drilling and Completion Fluids, Fifth ed. Gulf Publishing Company, Houston-USA p. 634.
12. Durie, R.W., Jessen, F.W., 1964. Mechanism of the dissolution of salt in the formation of underground Salt cavities. Soc. Pet. Eng. J. paper spe 678, 4 (2), 183–190.
13. Finneran, D.W., Morse, J.W., 2009. Calcite dissolution kinetics in saline waters. Chem. Geol. 268, 137–146.
14. Hamed, S.B., Belhadri, M., 2009. Rheological properties of biopolymers drilling fluids. J. Pet. Sci. Eng. 67, 84–90.
15. Hamida, T., Kuru, E., Pickard, M., 2009. Rheological characteristics of aqueous waxy hull-less barley (WHB) solutions. J. Pet. Sci. Eng. 69, 163–173.

16. Iscan, A.G., Kok, M.V., 2007. Effects of polymers and CMC concentration on rheological and fluid loss parameters of water-based drilling fluids. Energy Sources Part A 29, 939–949.
17. Luckham, P.F., Rossi, S., 1999. The colloidal and rheological properties of bentonite suspensions. Adv. Colloid. Interface Sci. 82 (1–3), 43–92.
18. McCabe, W.L., Smith, J.C., Harriott, P., 1985. Unit Operations of Chemical Engineering, Chemical Engineering Series, Fourth ed. McGraw-Hill International Editions, Singapore.
19. Menezes, R.R., Marques, L.N., Campos, L.A., Ferreira, H.S., Santana, L.N.L., Neves, G.A., 2010. Use of statistical design to study the influence of CMC on the rheological properties of bentonite dispersions for water-based drilling fluid. Appl. Clay Sci. 49, 13–20.
20. Meng, X., Zhang, Y., Zhou, F., Chu, P.K., 2012. Effects on carbon ash on rheological properties of water-based drilling fluids. J. Pet. Sci. Eng. 100, 1–8.
21. Morse, J.W., Arvidson, R.S., 2002. The dissolution kinetics of major sedimentary carbonate minerals. Earth-Sci. Rev. 58, 51–84.
22. Santos, M.F.L., Lana, P.C., Silva, J., Fachel, J.G., Pulgati, F.H., 2009. Effects of nonaqueous fluids cuttings discharge from exploratory drilling activities on the deep-sea macrobenthic communities. Deep-Sea Res. II 56, 32–40.
23. Sönmez, A., Kök, M.V., Özel, R., 2013. Performance analysis of drilling fluid liquid lubricants. J. Pet. Sci. Eng. 108, 64–73.

Chapter 10

Experimental Investigation of the Effect of Machining Parameters on the Surface Roughness and the Formation of Built Up Edge (BUE) in the Drilling of Al 5005

Erkan Bahçe[1] and Cihan Ozel[2]

[1]İnonu University, Department of Mechanical Engineering, Malatya, Turkey
[2]Firat University, Department of Mechanical Engineering, Elazig, Turkey

INTRODUCTION

Aluminum is one of the metals, besides iron and steel, which are widely used in many industrial fields such as aviation, navigation and automotive. The most considerable reasons why aluminum is so widely used are: i) it is light, ii) its alloys have higher strength than the construction steel, iii) it has high heat and electrical conductivity. Because of these excellent characteristics, the usage of aluminum as engineering material has an ever-increasing importance in several technological fields [1,2]. Although aluminum is widely used, there are many problems such as tool abrasion, burr formation and poor hole surface quality in drilling process of aluminum and its alloys [2,3]

The surface quality, which is one of these problems, is quite important for the efficient working of machine parts. The structure of a machined surface is one of the most important criteria in terms of quality, and tribological properties of the machined surface are considerably affected from the surface tissue. Generally the surface quality is characterized with surface roughness. Surface roughness is an important factor which must be considered not only in the conventional subjects of tribology such as abrasion, friction and lubrication but also in different fields such as sealing, hydrodynamics, electrical and heat conductivity. Surface roughness is mainly affected during the machining process by cutting parameters such as cutting speed, feed rate and depth of cut [4,5,6]. If these parameters are not chosen convenient, the surface roughness increases. This situation creates a notch effect and results in crack initiation, decrease in fatigue strength and corrosion resistance. So, the characterization and measurement of surface roughness has a great important in the sense of the optimization of machining process [7,8].

In former studies on hole surface quality, Nouari and his colleagues subjected Al 2024-T3 to dry drilling process, and did some optimizations and analysis experimentally for both the dimensional accuracy of the machined surface and longevity of cutting tools [9]. In their study, they used sintered tungsten carbide

(STC) cutting tools and high speed stell (HSS) cutting tools, and set the feed rate to 0,04 mm/rev and cutting speed to 25, 65, 165 m/min. They concluded from the experiments that STC cutting tools are more convenient in comparison with HSS cutting tools from the points of tool life, deviation in hole diameter and surface roughness. Lin investigated tool life, surface roughness, tool abrasion and burr formation for the process of the high speed machining of stainless steel material with TiN coated carbide tool [10]. As a result of his researches, he determined that the abrasions in shear edge result from the high feed rate in low cutting speed, and optimum cutting speed for desired burr height and surface roughness was 75 m/min. In addition, he determined that in high speed machining of stainless steels the tool life increased considerably in case of adjusting the feed rate to the values lower than 0,05 mm/rev. Lin and Syhu, studied on the treatment of the tool life and burr formation in the drilling of stainless steel with the drill bits coated by different materials [11]. Kurt et al., investigated the effect of cutting parameters on the drilling temperature, cutting force and surface roughness in the drilling of Al 2024 alloy with DLC coated drill. In their study, they determined that the most effective factors influencing the hole surface quality are feed rate and drill diameter [12]. They observed that the change in feed rate and diameter at high cutting speeds affects the average surface roughness considerably. Dudzinski et al., determined that the tool life was very short in the drilling of Inconel 718; therefore the surface quality gets worse [13]. They determined that the main wear mechanism seen in the cutting tools used was abrasion. In addition, they observed that the chips resulted in the formation of built-up-edge (BUE) by adhering on the cutting tool, and the removal of BUE from the cutting tool repeatedly caused notches. Kılıçkap investigated the roughness of hole surface and the height of the burrs formed at the hole exit in the drilling of Al 7075 material [14]. Also in another research, Kılıçkap, experimentally studied on the effects of cutting speed, feed rate and different cooling techniques on the temperature and the roughness of hole surface in the drilling of Al 7075 [15]. In their study, they observed that the most appropriate cooling technique was oil cooling from the point of good surface roughness. Also, they determined that the roughness increased with the increase of the feed rate, while it decreased with the increase of rotation speed. Hanyu et al. investigated the effects of finely crystallized diamond coating method, which was developed by themselves, on the surface roughness in the dry and semi-dry drilling of Al 7075 alloy [16]. They demonstrated experimentally that finely crystallized diamond coating method yields four times better results in comparison with the conventional diamond coating method. Konig and Grass investigated the effects of cutting parameters on the roughness of hole surface and surface tissue in the drilling of fiber reinforced thermosets [17]. They denoted that the surface roughness increases with increase of the feed rate. In his study, Tosun, optimized the drilling parameters affecting the burr height and surface roughness of DIN 42CrMo4 steel material by considering different drill materials, cutting speeds, drill point angles and feed rates with the help of Grey Relational Analysis (GRA) [18]. Sur et al. studied on the effects of Ti alloy on the surface roughness in the turning of Al 6063 alloy [19]. They observed that the increase of 35 percent in the hardness of the material resulting from the doping of Ti to the material had relatively an inconsiderable effect on the surface roughness of the material in comparison with effects of cutting speed and feed rate. Also, they determined that the increase in the feed rate affected the surface roughness negatively, while the increase in the cutting speed contributed to the treatment of surface roughness positively; however the feed rate had a more dominant effect on the surface roughness in comparison with the cutting speed. Darwish, et al., investigated the effects of cutting speed, feed rate and drill diameter on the hole surface quality, dimensional accuracy and geometric tolerance in soft steel materials [20]. In their study, they observed that cutting speed and feed rate had a great effect on surface quality, and the higher dimensional accuracy was obtained at low cutting speeds and feed rates.

In the studies mentioned above, generally the effects of cutting parameters on the roughness of the hole surface were investigated in the machining process of stainless steel and 2000, 6000 and 7000 series aluminum alloys. However, it has drawn attention that the studies on 5000 series aluminum alloys, which are widely used in many industrial fields such as aviation, navigation and automotive, are not sufficient. In this study, Al 5005 material was drilled on CNC milling machine under dry drilling conditions by considering different machining parameters such as various rotation speeds, feed rates, drill diameters and point angles, and the roughness of hole surface and the formation of BUE on cutting edges were investigated.

EXPERIMENTAL METHOD

In this study, Al 5005 was drilled by considering various drilling parameters such as diameter, point angle, feed rate and rotation speed. CNC milling machine (Taksan, TMC 700V) with vertical machining centre was used in the experiments. The spindle power of the machine, rotation speed and feed rate values were taken as 5.5 kW, 50-8000 rev/min and maximum 0.6 mm/rev, respectively. Maximum feed rate values of the work table on X, Y and Z axes were 500, 600 and 450 mm, respectively. Factorial design, in which the effects of mostly different and unrelated factors on a definite characteristic are investigated, was taken into consideration in design process of the experiment. In factorial design, the experimental design is established by processing the variable parameters (or their levels) crossingly [21]. In this study, the experiments were conducted in accordance with 72 different combinations ($2^1.3^2.4^1$) by using 2 levels for the drill diameter, 4 levels for the point angle, 3 levels for the rotation speed and the feed rate. The values of variable parameters in conducted experiment were selected in compliance with the similar studies as shown in Table 1 [9,10,14,18].

In this study, the cutting fluid was not used in order to observe the effect of drill parameters on the roughness of the hole surface [22]. Al 5005 material used in the experiments was in the dimension of 10mmx70mmx400mm, and its chemical properties were given in Table 2. In the drilling process, the space between the axes of each hole on the sample was adjusted to be 20 mm (Figure 1).

Table 1: Experimental parameters

Feed Rate (mm/rev)	Rotation Speed (rev/min)	Point Angle (degree)	Drill material and diameter (mm)
0.1, 0.2, 0.3	400, 800, 1200	90, 118, 130, 140	HSS, Ø5, Ø10

Table 2: The chemical structure of Al 5005

Al 5005	Mg	Si	Fe	Cu	Mn	Cr	Zn	Other elements	Al
%	0.5-1.10	0.3	0.3	0.20	0.20	0.10	0.25	0.15	remainder

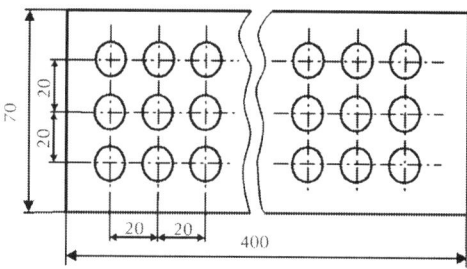

Figure 1: The space between the drilling axes (Thickness 10 mm).

N type double-end DIN 338/RN HSS drill bits with 30° helix angle were used in drilling process. Hardness value of these cutting tools was 65 HRc. Each cutting tool was used once in the experiments, and each experiment was repeated three times in accordance with the similar studies [9,11-14,21,22]. The images of BUEs formed on the cutting tool as a result of the drilling processes were taken by means of Leo Evo 40 model Scanning Electron Microscope (SEM).

After combinational drilling processes, the samples were cut with a cutting disc in the middle in parallel with the hole axis in order to measure the roughness of the hole surface. Then, the surface roughnesses were measured with Mitutoyo SJ-201 surface roughness measurement device. In the measurement of the roughness, sampling length and sampling number were chosen by considering the former studies [7,8,16,19,21] as 0.8 mm and 5 (0.8x5), respectively. The other sampling length values of this device were 0.25 mm and 2.5 mm. Generally, the roughness was measured at three different points in parallel with the hole axis in accordance with the studies in literature [7,12,16,18]. But in this study, in order to evaluate the measurements accurately, the measurements were taken from 5 different points, and then Ra values were determined by considering the average of these values.

RESULTS AND DISCUSSION

The graphics in Figure 2 were illustrated to enable one a comprehensive assessment of the effects of drilling parameters on the surface roughness in the drilling of Al 5005 without using cooling fluid.

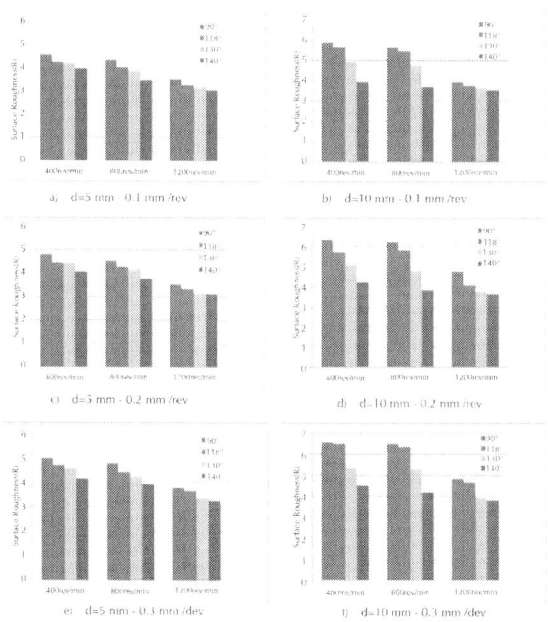

Figure 2: The change of the surface roughness with the drilling param-

eters.

As seen from the graphics, the surface roughness decreases with the increase of rotation speed. In former studies [18,23], this case was attributed to the decrease in the cutting and feed force. The most considerable reasons of the decrease in these forces are the decrease in the contact region between the tool and work piece and the decrease in the shear strength in the cutting region due to the increase in the heat of tool-work piece depending on the cutting speed [18,24].

Also, it was supposed that the influence of BUE on the tool and material increased negatively since the amount of BUE resulting from the adhesive wear increased due to the increase in the feed and cutting forces. As mentioned in the literature, in the machining processes of many alloys including more than one phase in own structures BUE formed due to the adhesion of the chips on the tool surface and cutting edge because of the work hardening [23]. It can be said that BUE especially forming at low cutting speeds affects the surface roughness negatively. since Aluminum and its alloys includes more than one phase. In order to explain this case clearly, BUE formations occurring on the cutting tool edges were also investigated in this study (Figures 3 and 4 a-c). All the images corresponding to the whole cutting parameters were not presented since there were a lot of parameters in the experiment and they were investigated in other sections separately, but the SEM images expressing the mentioned case clearly were presented.

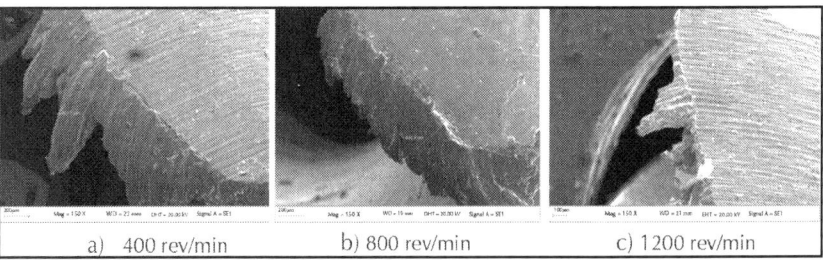

a) 400 rev/min   b) 800 rev/min   c) 1200 rev/min

Figure 3: SEM images of BUE formation on the cutting edges. (0.2

mm/dev-118°- Ø5 mm).

a) 400 rev/min   b) 800 rev/min   c) 1200 rev/min

Figure 4: SEM images of BUE formation on the cutting edges. (0.1 mm/rev-118°-Ø10 mm).

As seen from the SEM images in Figures 3 and 4 a-c, BUE formation decreased and had a minor effect on the surface roughness because of the increase in the rotation speed, so surface roughness decreased (Figures 4a-f). Since BUE formed on the cutting tool edge during the drilling had an unstable structure, the surface roughness increased. Thus, because of big and unstable BUE due to low cutting speeds (Figures 3, 4 a and b), the surface roughness increased further and a bad surface was formed (Figures 2a-f). The decrease in BUE due to the increase in cutting speed can be ascribed to the increase in temperature [23,25]. Since high cutting speeds resulted in much more increase in the temperature, BUE on the cutting edge lost its hardness and strength, and in the continuing cutting process it couldn't resist the tensions on itself and it was removed from the cutting edge (Figure 3 and 4 c). Hence, high cutting speeds reduced the tendency to the formation of BUE, and resulted in the decrease in the surface roughness values of the work piece (Figures 2 a-f). Since BUE formed on the cutting edges also spoiled the geometric structure of the cutting tool, the stable and ideal process of the cutting operation was damaged, so the roughness increased (Figure 2 a-f). Also, BUE formed on the cutting edges caused fracturing and abrasion on the cutting edges while it was separating from the cutting edges by the effect of thermal tensions [24]. This case increased the roughness of the hole surface depending on the size of BUE (Figures 4 a-c). The abrasion and fracturing formed on the cutting edge according to the different machining parameters were presented in Figures 5 and 6, respectively.

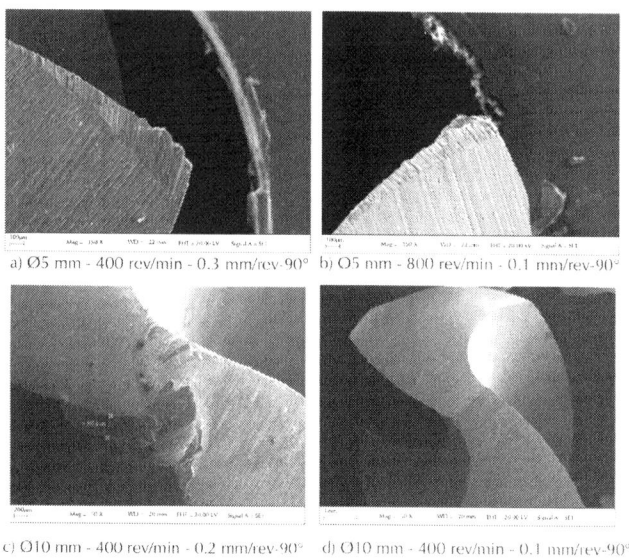

a) Ø5 mm - 400 rev/min - 0.3 mm/rev-90° b) Ø5 mm - 800 rev/min - 0.1 mm/rev-90°

c) Ø10 mm - 400 rev/min - 0.2 mm/rev-90° d) Ø10 mm - 400 rev/min - 0.1 mm/rev-90°

Figure 5: SEM images of the abrasion formed on the cutting edges.

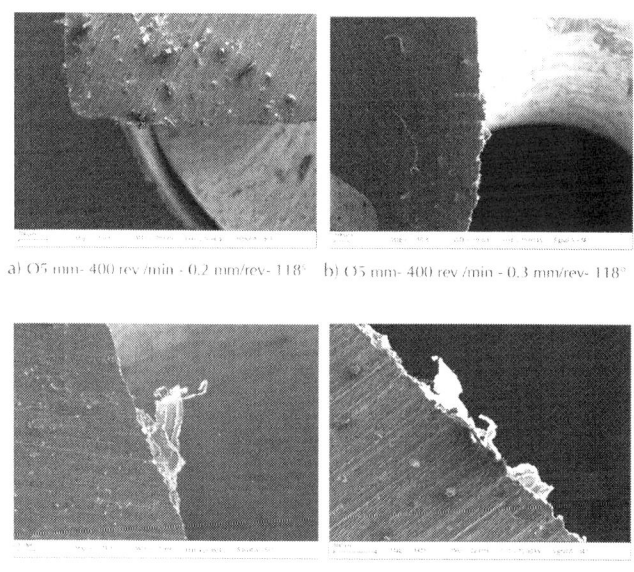

a) Ø5 mm- 400 rev/min - 0.2 mm/rev- 118° b) Ø5 mm- 400 rev/min - 0.3 mm/rev- 118°

c) Ø10 mm- 1200 rev/min - 0.3 mm/rev- 118° d) Ø10 mm- 400 rev/min - 0.2 mm/rev- 130°

Figure 6: SEM images of the fractures formed on the cutting edges.

In a similar manner as BUE formation, the chips adhering to helical channels obstructed the effective removal of the chips by plugging the helical channels partially (Figure 7). This case was more prominent at low rotation speeds, and the roughness increased depending on the this case (Figures 2 a-f).

As seen from the Figure 2 a-f, surface roughness increased with the increase of the feed rate, since the amount of the chips increased due to the increase in the feed rate. Because the increase of the feed rate caused high feed rate, low shear angle and thick chip formation [26]. This case signified that machined surfaces were more influenced from the forces during the cutting process. Likewise, the increase in the feed rate resulted in high friction resistance, pressure and increase in temperature [23,25]. In this case, the chip experienced to shear tensions, and adhered to the cutting tool. The amount of the chip adhesion increased depending on the feed rate, and an unstable structure formed. In order to explain this case better, chip smearing formed on the cutting edge were also investigated (Figure 8). It was supposed that notch effect of these chips on the machined due to the adhesion between the chip and cutting edge caused a corruption on the surface quality (Figure 8). This case resulted in the increase in surface roughness.

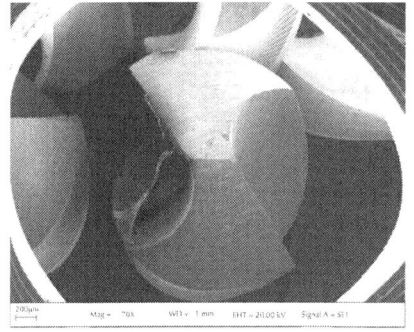
a) Ø5 mm- 400 rev /min- 0.2 mm/rev- 90°

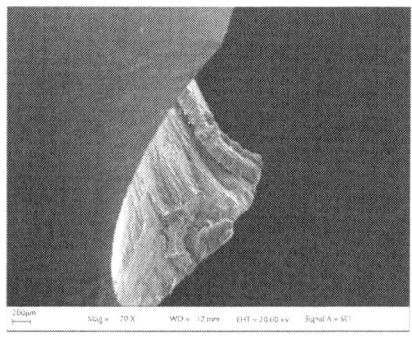
b) Ø5 mm- 400 rev /min- 0.3 mm/rev- 90°

Figure 7: SEM images of the chips adhering to the helical channels.

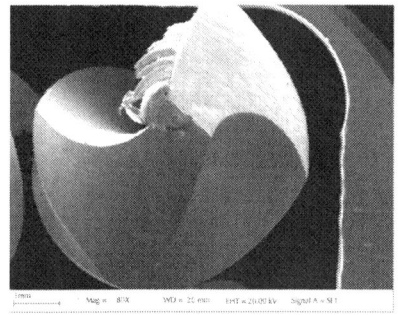
90°-0.1 mm /rev-800 rev/min

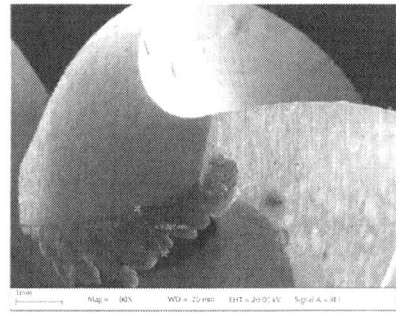
90°-0.3 mm /rev-400 rev/min

Figure 8: SEM images of the smearing formed on the cutting tool.

In addition, surface roughness changes depending on the feed rate and cutting radius in turning as denoted Ra=0.321(f^2/r) in the Ref. [27]. Where R_a is the roughness, f is the feed rate, r is the radius.

Monagham and O'Reily, determined that this relation was valid for the drilling process, and a similar relation emerged in the drilling of Al 5005 (Figure 2 a-f).

Similarly, as seen from Figure 2 a-f surface roughness improved with the increase of drill point angle. This case can be explained as follows: The values of plastic deformation region, cutting edge length and chip thickness obtained for the drills having point angles of 118°, 130° and 140° are greater than those obtained for the drill having point angle of 90°. Furthermore, since the cutting tool was worn away faster due to the expansion of the friction surface of the cutting edge with the decrease of the point angle [28] the stability of the cutting process was influenced negatively, therefore the roughness increased as seen from the Figure 2 a-f. Also, the pressure applied on the hole surface was decreased owing to the decrease in radial force with the increase of the point angle. Hence, the roughness arisen on the surface was less in comparison with the drills having small point angles.

On the other hand, it was supposed that surface roughness was also influenced by BUE arisen on the cutting edge. The change in BUE on the cutting tool depending on the different point angles was illustrated in Figure 9. As seen, as the point angle decreased, BUE influenced the form of the cutting tool more negatively. Therefore, this case affected the stability of the cutting tool during the cutting process, and caused an increase in surface roughness (Figure 2 a-f).

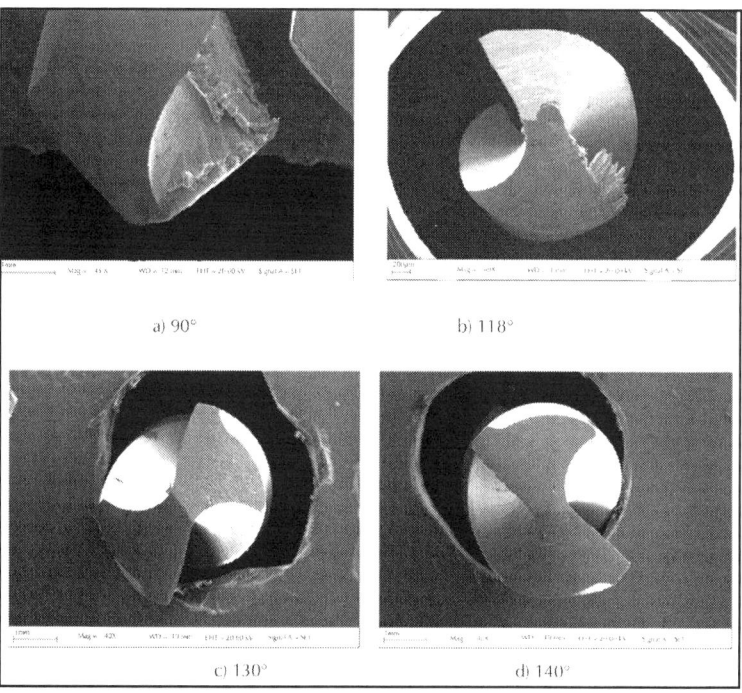

Figure 9: SEM images of BUE formation on the cutting edges at different point angles (1200 rev/min-0.1 mm/rev).

In addition, it was seen in the drilling process that the roughness values obtained for a drill with a diameter of Ø10 mm was bigger than those obtained for a drill with a diameter of Ø5 mm (Figure 2 a-f). This case was attributed to the increase of the forces due to the increase of the length of the cutting edge [23,28]. Also, as seen from Figure 10, it was supposed that the expansion of the deformation area due to the increase of drill diameter resulted in a rise in BUE formation which caused an increase in the roughness (Figure 2 a-f).

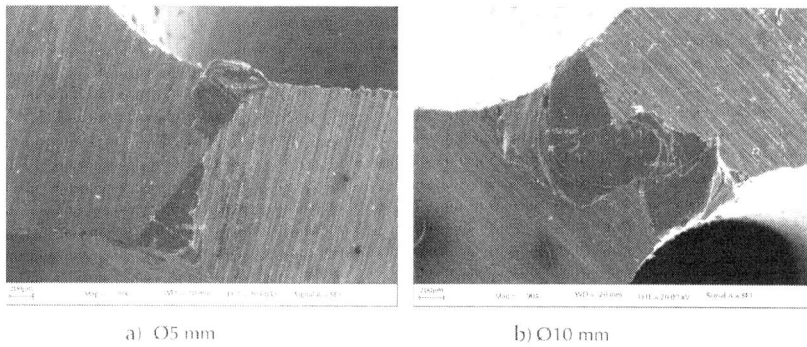

a) O5 mm b) O10 mm

Figure 10: SEM images of BUE formation on the cutting edges at different drill diameters. (800 rev/min-0.2 mm/rev-118°).

CONCLUSIONS

In this study, BUEs arisen on the cutting edges and the effect of drilling parameters (rotation speed, feed rate, drill diameter and point angle) on the surface roughness of the work piece were investigated experimentally in the drilling process of Al 5005 alloy on CNC milling machine. The inferences achieved were presented as follows:

- The surface roughness decreased with the increase of the rotation speed and point angle, while it increased with the increase of the feed rate and drill diameter.
- It was observed that BUE arisen on the cutting edges caused fractures and wears on the cutting edges during the removal from the cutting edges.
- BUE formation decreased with the increase of the rotation speed, therefore the value of surface roughness decreased.
- BUE plugged the helical channels partially by adhering on them, and obstructed the effective removal of the chips.
- BUE spoiled the form of the cutting tool more as the drill point angle decreased, and this case resulteZd in an increase in the surface roughness.

REFERENCES

1. W. S. Smith, 2001translation: Mehmet Erdo an. Structure and Properties of Engineering Alloys. 2
2. P Stringer, ., G Byrne, . ve Ahearne, E., 2010. "Tool design for burr removal in drilling operations",http://www.ucd.ie/mecheng/ams/news_items/Peter%20Stringer. pdf, 17
3. J. Kim, ve Dornfeld, D. A., 2002Development of an Analytical Model for Drilling Burr Formation in Ductile Materials", Transactions of the ASME, 124
4. I. Puertas, Luis perez, C.J., 2003Surface rougness prediction by factorial design of experiments in turning processes", Journal of Materials Processing Technology 143144
5. C. Çogun, B. Özses, 2002Effect Of Machining Parameters On Surface Roughness In Cnc Machine Tools", Gazi Üniv.Müh. Mim.Fak. Der. 171
6. J. Koelsch, 2001Divining Edge Quality by Reading the Burrs", Quality Magazine, December, 2428
7. D. Karayel, 2008Prediction and control of surface roughness in CNC lathe using artificial neural networkJournal of Materials Processing Technology, 20931253137
8. M. H. Çetin, B. Özçelik, E. Kuram, B. T. Şimşek, E. Demirbeş, 2010Effect of Feed Rate on Surface Roughness And Cutting Force In The Turning of AISI 3041 Steel With Ep Added Vegetable Based Cutting Fluids", Selçuk University, II. National Machining Symposium, UTİS 2010, S:92107Konya, Turkey
9. M Nouari, .; G List, .; F Girot, . & D Gehin, . 2005, Effect of machining parameters and coating on wear mechanisms in dry drilling of aluminium alloys. International Journal of Machine Tools & Manufacture, 45 12-13 , 14361442 , 0890-6955
10. T. R. Lin, 2002Cutting behavior using variable feed and variable speed when drilling stainless steel with TiN-coated

carbide drills. International J. Advance Manufacturing Technology. 19629636

11. T. R. Lin, ve Shyu, R. F., 2000mprovement of Tool Life and Exit Burr using Variable Feeds when Drilling Stainless Steel with Coated Drills, Int. J. Adv. Manuf. Technol, 16308313
12. M. Kurt, Y. Kaynak, B. Bakir, U. Köklü, G. Atakök, L. Kutlu, 2009Experimental Investigation And Taguchi Optımızatıon For The Effect Of Cuttıng Parameters On The Drilling Of Al 2024T4 Alloy With Diamond Like Carbon (DLC) Coated Drills", 5. Sixth International Advanced Technologies Symposium (IATS'09), Karabük, Turkey
13. D. Dudzinski, A. Devillez, A. Moufki, D. Larrouquere, V. Zerrouki, J. Vigneau, 2004A review of developments towards dry and high speed machining of Inconel 718 alloyInt. J. Mach. Tools Manufact. 44 (4), 439 EOF456 EOF
14. E Kılıckap, ., 2010, "Modeling and optimization of burr height in drilling of Al-7075 using Taguchi method and response surface methodology", Int J Adv Manuf Technol, DOI 10.1007/s00170-009-2469-x.
15. E. Kiliçkap, 2009Investigation of The Effect of Cutting Parameters on The Burr Formatıon In Drıllıng of Al-7075', I. National Machining Symposium, 142150Istanbul, Turkey
16. H. Hanyu, S. Kamiya, Y. Murakami, M. Saka, 2003Dry and semi-dry machining using finely crystallized coating cutting tools" Surface and Coatings Technology 173174
17. W. König, and P. Grass, 1989Quality Definition and Assessment in Drilling of Fibre Reinforced ThermosetsCIRP Annals, 381119124
18. N. Tosun, 2006Determination of optimum parameters for multi-performance characteristics, in drilling by using grey relational analysis", Int J Adv Manuf Technol 28450455
19. G. Sur, H. Çetin, E. Çevik, H. Ahlatçi, Y. Sun, 2011Determining the Influence of Ti Additive on Surface Roughness During Turning of AA6063 Alloy", 6th International Advanced Technologies Symposium (IATS'11), Elazı , Turkey.

20. S. M. Darwish, A. , M. Tamitni, 1997Formulation of Surface Roughness Models for Machining Nickel Super Alloy with Different ToolsMaterials and Manufacturing ProcessesVol.12, 3395 EOF
21. N. Tosun, C. Kuru, E. Altinta , ve Erdinç E., 2010Investıgatıon of Surface Roughness In Milling With Air And Conventional Cooling Method", J. Fac. Eng. Arch. Gazi Univ. 251
22. T. Obikawa, Y. Kamata, J. Shinozuka, 2006High-speed grooving with applying MQLInternational Journal of Machine Tools & Manufacture, 46, 1854 EOF1861 EOF
23. Y. ahin, 2000Principles of Metal Removing", 1-2Nobel Publisher.
24. E. Kilickap, M. Huseyinoglu, and C. Ozel, Emprical Study Regarding the Effercts of Minimum Quantity Lubricant Utilization on Performance Characteristics in the Drilling of Al 7075", J. Of the Braz. Soc. Of Mech. Sci.&Eng. Vol.XXXIII, 1
25. B. Özcelik, ve Bagci, E., 2006Experimental and numerical studies on the determination of twist drill temperature in dry drilling": A new approach, P I Mech Eng L-J Mat, 27920927
26. C. Çakir, The Fundamentals of Modern Machining", 1999Uluda University, Bursa, Turkey
27. J. Monagham, O. Reily, 1992The drilling of an Al/SiC metal matrix composite", J Mater Process Tech, 33, 469-480.
28. M. Akkurt, 2004Metal Removing Methods And Machine", Birsen Publisher, Istanbul, Turkey

Citations

CHAPTER 1

Titus N. Ofei, Sonny Irawan, and William Pao, "CFD Method for Predicting Annular Pressure Losses and Cuttings Concentration in Eccentric Horizontal Wells," Journal of Petroleum Engineering, vol. 2014, Article ID 486423, 16 pages, 2014. doi:10.1155/2014/486423.

CHAPTER 2

S. Abbasi, S. Zebarjad and S. Baghban, "Decorating and Filling of Multi-Walled Carbon Nanotubes with TiO_2 Nanoparticles via Wet Chemical Method," Engineering, Vol. 5 No. 2, 2013, pp. 207-212. doi:10.4236/eng.2013.52030.

CHAPTER 3

Makhkhas, Y. , Aqdim, S. and Sayouty, E. (2013) Study of Sodium-Chromium-Iron-Phosphate Glass by XRD, IR, Chemical Durability and SEM. Journal of Materials Science and Chemical Engineering, 1, 1-6. doi:10.4236/msce.2013.13001.

CHAPTER 4

F.L. Zhang, P. Liu, L.P. Nie, Y.M. Zhou, H.P. Huang, S.H. Wu, H.T. Lin, A comparison on core drilling of silicon carbide and alumina engineering ceramics with mono-layer brazed diamond tool using surfactant as coolant, Ceramics International, Available online 9 April 2015, ISSN 0272-8842, http://dx.doi.org/10.1016/j.ceramint.2015.03.117.

CHAPTER 5

Jana D. Abou Ziki, Rolf Wüthrich, Nature of drilling forces during spark assisted chemical engraving, Manufacturing Letters, Volume 4, April 2015, Pages 10-13, ISSN 2213-8463, http://dx.doi.org/10.1016/j.mfglet.2015.01.001.

CHAPTER 6

Michael Nikolaou, Pratik Misra, Vincent H. Tam, Andrew D. Bailey III, Complexity in semiconductor manufacturing, activity of antimicrobial agents, and drilling of hydrocarbon wells: Common themes and case studies, Computers & Chemical Engineering, Volume 29, Issues 11–12, 15 October 2005, Pages 2266-2289, ISSN 0098-1354, http://dx.doi.org/10.1016/j.compchemeng.2005.05.028.

CHAPTER 7

Hiroki Endo, Etsuo Marui, Small-hole drilling in engineering plastics sheet and its accuracy estimation, International Journal of Machine Tools and Manufacture, Volume 46, Issue 6, May 2006, Pages 575-579, ISSN 0890-6955, http://dx.doi.org/10.1016/j.ijmachtools.2005.07.026.

CHAPTER 8

Q.H Zhang, J.H Zhang, D.M Sun, G.D Wang, Study on the diamond tool drilling of engineering ceramics, Journal of Materials Processing Technology, Volume 122, Issues 2–3, 28 March 2002, Pages 232-236, ISSN 0924-0136, http://dx.doi.org/10.1016/S0924-0136(02)00016-X.

CHAPTER 9

L.A. Calcada, L.A.A. Martins, C.M. Scheid, S.C. Magalhães, A.L. Martins, Mathematical model of dissolution of particles of NaCl in well drilling: Determination of mass transfer convective coefficient, Journal of Petroleum Science and Engineering, Volume 126, February 2015, Pages 97-104, ISSN 0920-4105, http://dx.doi.org/10.1016/j.petrol.2014.12.011.

CHAPTER 10

Erkan Bahçe and Cihan Ozel (2013). Experimental Investigation of the Effect of Machining Parameters on the Surface Roughness and the Formation of Built Up Edge (BUE) in the Drilling of Al 5005, Tribology in Engineering, Dr. Hasim Pihtili (Ed.), ISBN: 978-953-51-1126-9, InTech, DOI: 10.5772/56027.

Index

A
Abrasive water jet machining (AWJM) 72

B
Built-up-edge (BUE) 227

D
Drilling fluid 204, 223
Drilling forces 91, 92, 99, 244
Drilling machine 174, 175

E
Electrical discharge machining (EDM) 72

F
Full width at half maximum intensity (FWHM) 43

G
Gather momentum 188
Grey Relational Analysis (GRA) 228

H

hHgh speed stell (HSS) 227

K

Karhunen–Loeve (KL) 108

L

Laser beam machining (LBM) 72
Linear path 192

M

Material removal rate 188, 189, 193, 195, 197, 198, 199
Multi-walled carbon nanotubes (MWCNTs) 39

N

Natural science foundation of china (NSFC) 86

O

Ordinary differential equation (ODE's) 202

P

Partial differential equations (PDE's) 202
Partial Differential Equations (PDE's) 204
Prediction 2
Principal component analysis (PCA) 109
Pyrophosphate 56, 58, 61

R

Rate of penetration (ROP) 3, 10

Response surface methodology (RSM) 107
Root mean square (RMS) 16
Rotary diamond tool 189
Rotational speed 188, 189, 193, 195, 196, 197, 198, 199

S

Scanning Electronic Microscopy (SEM), 56
Scanning Electron Microscope (SEM) 230
Sintered tungsten carbide (STC) 227
Sodium-chromium-iron phosphate glass 56, 61
Spark assisted chemical engraving (SACE) 92

T

Technical Support Unit for Scientific Research (TSUSR) 65
Tool–glass detachment 97
Transmission electron microscopy (TEM) 40

W

Water-based muds (WBM) 202, 205
Water-based mud (WBM) 203

X

X-ray diffraction (XRD) 40
X-Ray Diffraction (XRD), 56